Susanne Schmitt

Detection and Characterization of Inclusions in Impedance Tomography

Detection and Characterization of Inclusions in Impedance Tomography

by
Susanne Schmitt

Dissertation, Karlsruher Institut für Technologie
Fakultät für Mathematik
Tag der mündlichen Prüfung: 22.12.2010
Referenten: Prof. Dr. Andreas Kirsch, PD Dr. Tilo Arens

Impressum

Karlsruher Institut für Technologie (KIT)
KIT Scientific Publishing
Straße am Forum 2
D-76131 Karlsruhe
www.ksp.kit.edu

KIT – Universität des Landes Baden-Württemberg und nationales
Forschungszentrum in der Helmholtz-Gemeinschaft

KIT Scientific Publishing 2011
Print on Demand

ISBN 978-3-86644-635-9

Preface

The topic of this thesis are two further developments of the Factorization method for electrical impedance tomography (EIT). In EIT current is applied to the surface of the investigated subject and the resulting voltage is measured at the surface. From these measurements one tries to recover information about the conductivity inside the subject. The Factorization method for EIT is a noniterative method to detect domains inside the investigated subject that exhibit a different conductivity than the a priori known background medium.

We start by giving an introduction to the direct and the inverse problem of EIT in Chapter 1. We show how our mathematical model can be derived from Maxwell's equations and give an outline of the functional analytic setting we are dealing with. Afterwards we discuss the direct and the inverse problem of EIT and give a short summary of some reconstruction methods and in particular the Factorization method.

For the Factorization method for EIT it is usually assumed that either all the inclusions have a higher or they all have a lower conductivity than the background medium. In Chapter 2 we therefore develop a modification of the Factorization method for EIT that is capable of detecting mixed inclusions, i.e. in the case in which there are both inclusions with a higher as well as inclusions with a lower conductivity than the background medium. Parts of Chapter 2 have been previously published in the paper [71].

Since the Factorization method only provides information about shape and location of inclusions but not about their actual conductivity we present a method to compute the conductivity inside inclusions after they have been localized in Chapter 3. This method is based on a new version of the Factorization method for EIT that involves a factorization with three operators that are different from those in Chapter 2. In particular, we show some essential properties of the spectrum of the operator that appears in the middle of this new factorization and that it is closely related to the conductivity of the inclusions.

This work has partly been supported by the German Federal Ministry of Education an Research (BMBF) under the project 'Regularization Methods for Electrical Impedance Tomography in Medicine and Geoscience '. This financial support is gratefully acknowledged.

Furthermore, this work would not exist without the support of my colleagues at the department of mathematics of the Karlsruhe Institute of Technology. First of all, I would like to thank my advisor Prof. Dr. Andreas Kirsch for many fruitful discussions as well as the excellent supervision during the recent years. I also thank PD Dr. Tilo Arens for being the co-examiner of this thesis and for a lot of encouragement from the beginning of my work on. Moreover, I am much obliged to Dr. Armin Lechleiter, Andreas Helfrich-Schkarbanenko and Sven Heumann for many stimulating discussions and valuable remarks. Finally, I would like to thank PD Dr. Frank Hettlich, Marc Mitschele, Dr. Kai Sandfort, Monika Behrens, Dr. Karsten Kremer, Dr. Sebastian Ritterbusch and Dr. Slavyana Geninska for their help and for providing a very friendly working atmosphere.

Contents

1 Electrical Impedance Tomography

This first chapter serves as an introduction to this work. After a short summary of possible technical applications of electrical impedance tomography we turn towards the investigation of the direct problem. Afterwards we formulate the inverse problem and present some important results concerning identifiability and various reconstruction methods. At last we give an outline of the development of the Factorization method and explain what new results our work contributes to the Factorization method for impedance tomography.

1.1 Motivation

In electrical impedance tomography (EIT) current is applied to the surface of the investigated subject and the resulting electrical potential is measured at the surface. From a set of such measurements one tries to obtain information about the conductivity inside the subject.

There is a large variety of possible applications of this imaging method. First of all, there are applications in medicine such as lung imaging or the detection of breast tumors. Since different tissues inside the body have different conductivities, these tissues can potentially be visualized using EIT. In contrast to other imaging methods in medicine, production as well as application of EIT devices are relatively cheap and there are no harmful side-effects such as radiation exposure as they are unavoidable e.g. for X-ray tomography.

Another important field of application is geoelectrical imaging where one tries to recover information about the conductivity distribution in the ground. Since different materials in the ground exhibit different electrical properties, they can also be distinguished using EIT.

In the following section we derive the mathematical model for the direct problem of EIT, and afterwards we formulate the corresponding inverse problem and give an outline of some reconstruction methods.

1.2 The Direct Problem

In this section we investigate the direct problem of electrical impedance tomography. At first we give an outline how the mathematical model can be derived from Maxwell's equations. Afterwards we explain our functional analytic setting to ensure unique solvability and introduce the Neumann-to-Dirichlet operator.

1.2.1 Derivation of the Mathematical Model

We first assume that the investigated subject is three-dimensional while later we will see that the resulting direct problem for EIT applies to subjects in $\mathbb{R}^2$ as well. Let B be the subject unter investigation, i.e. $B \subset \mathbb{R}^3$ is a bounded and simply connected C^2-domain. The following considerations are adopted from [16].

The starting point for the derivation of the mathematical model for EIT are Maxwell's equations for conductive materials in the frequency domain:

$$\text{curl} E = i\omega\mu H, \quad \text{curl} H = (\sigma - i\omega\epsilon)\, E. \tag{1.1}$$

Let $[x], [E], [H] \in \mathbb{R}$ be scaling factors such that $x = [x]\,\tilde{x}$, $E(x) = E([x]\,\tilde{x}) = [E]\,\tilde{E}(\tilde{x})$ and $H(x) = H([x]\,\tilde{x}) = [H]\,\tilde{H}(\tilde{x})$. In particular, the terms in brackets carry the corresponding units while the quantities with tildes don't. Now we obtain

$$\text{curl}_{\tilde{x}}\tilde{E} = i\omega\mu\frac{[H]\,[x]}{[E]}\tilde{H}, \quad \text{curl}_{\tilde{x}}\tilde{H} = (\sigma - i\omega\epsilon)\,\frac{[E]\,[x]}{[H]}\tilde{E}.$$

Here $\text{curl}_{\tilde{x}}\tilde{E}$ is defined by $\text{curl}_{\tilde{x}}\tilde{E}(\tilde{x}) = \text{curl}\frac{[E]}{[x]}E(x)$ and, analogously, $\text{curl}_{\tilde{x}}\tilde{H}(\tilde{x})$ denotes $\text{curl}\frac{[H]}{[x]}H(x)$.

The mean value of σ in B is denoted by $\bar{\sigma}$, and we choose the scaling factors such that $\bar{\sigma}\frac{[E][x]}{[H]} = 1$. Hence the equations can be transformed to

$$\text{curl}_{\tilde{x}}\tilde{E} = i\omega\mu\bar{\sigma}\,[x]^2\,\tilde{H}, \quad \text{curl}_{\tilde{x}}\tilde{H} = \frac{1}{\bar{\sigma}}\,(\sigma - i\omega\epsilon)\,\tilde{E}.$$

Now the complex-valued admittivity is defined by $\gamma := \frac{1}{\bar{\sigma}}\,(\sigma - i\omega\epsilon)$ from which it is easy to see that $\text{Re}\,\gamma$ is non-negative and has mean value 1 while $\text{Im}\,\gamma$ is non-positive and depends on the applied frequency ω. In this work the complex-valued admittivity γ will be called (complex-valued) conductivity.

For non-ferromagnetic materials and low frequencies ω as they are usually used in EIT the term $\omega\mu\bar{\sigma}\,[x]^2$ is negligible. We therefore postulate $\text{curl}_{\tilde{x}}\tilde{E} = 0$ in B which also implies $\text{curl}E = 0$ in B and thus that there exists an electrical potential u such that $E = \nabla u$. Plugging this equality into the second equation in (1.1) and applying the divergence yields $\text{div}\,(\gamma\nabla u) = 0$.

The current applied to the boundary ∂B can be modelled by a Neumann boundary condition. To derive this boundary condition from Maxwell's equations we have to add the current density J to right hand side of (1.1) to obtain

$$\text{curl}H = (\sigma - i\omega\epsilon)\,E + J. \tag{1.2}$$

By substituting $E = \nabla u$ in (1.2) and using the divergence theorem in small domains at the boundary ∂B the Neumann boundary condition $\gamma\frac{\partial}{\partial v}u = f$ for $f = v \cdot J$ at ∂B can be obtained. For a more detailed derivation we refer to [16].

On the whole, we obtain a boundary value problem for the electrical potential u:

$$\text{div}\,(\gamma\nabla u) = 0 \text{ in } B, \quad \gamma\frac{\partial}{\partial v}u = f \text{ on } \partial B,$$

and in order to analyse this problem we present some required functional analytic tools in the following subsection.

1.2.2 Functional Analytic Setting

We proceed by giving an outline of the the functional analytic setting of this work. For more extensive introductions we refer to [66], [21] or [29]. Let B be a bounded domain in $\mathbb{R}^d$ for $d \in \{2, 3\}$ with C^2-boundary.

First of all we will deal with the following function spaces on the boundary:

$$C^2_\diamond(\partial B) = \{g \in C^2(\partial B) : \langle g, 1 \rangle = 0\},$$
$$L^2_\diamond(\partial B) = \{g \in L^2(\partial B) : \langle g, 1 \rangle = 0\},$$

where $\langle \cdot, \cdot \rangle$ denotes the standard scalar product in L^2. It is obvious that $C^2_\diamond(\partial B)$ is a dense subspace of $L^2_\diamond(\partial B)$. Now we consider the Sobolev space $H^{\frac{1}{2}}(\partial B)$ and its dual space, $H^{-\frac{1}{2}}(\partial B)$. It is well-known that the imbedding $j : H^{\frac{1}{2}}(\partial B) \to L^2(\partial B)$ is compact and has dense range (see e.g. Thm. 3.27 in [66]).

Using the Riesz representation theorem we identify $L^2(\partial B)$ with its dual space which implies that the scalar product in $L^2(\partial B)$ is identical to the dual evaluation between $L^2(\partial B)$ and its dual space. Furhermore, this dual form extends to the dual pairing in $H^{-\frac{1}{2}}(\partial B) \times H^{\frac{1}{2}}(\partial B)$.

Now we define two more function spaces on ∂B:

$$H^{\frac{1}{2}}_\diamond(\partial B) = \{g \in H^{\frac{1}{2}}(\partial B) : \langle g, 1 \rangle = 0\},$$
$$H^{-\frac{1}{2}}_\diamond(\partial B) = \{g \in H^{-\frac{1}{2}}(\partial B) : \langle g, 1 \rangle = 0\}.$$

Here, in the first line $\langle \cdot, \cdot \rangle$ denotes the scalar product in $L^2(\partial B)$, while in the second line it is meant in the sense of the dual evaluation between $H^{-\frac{1}{2}}(\partial B)$ and $H^{\frac{1}{2}}(\partial B)$. These two spaces are closed subspaces of $H^{\frac{1}{2}}(\partial B)$ and of $H^{-\frac{1}{2}}(\partial B)$, respectively. Now we can argue as before and identify $L^2_\diamond(\partial B)$ with its dual space. We observe that $H^{\frac{1}{2}}_\diamond(\partial B)$ is compactly embedded in $L^2_\diamond(\partial B)$ and that $H^{-\frac{1}{2}}_\diamond(\partial B)$ is the dual space of $H^{\frac{1}{2}}_\diamond(\partial B)$. This leads to the Gelfand triple

$$H^{\frac{1}{2}}_\diamond(\partial B) \subset L^2_\diamond(\partial B) \subset H^{-\frac{1}{2}}_\diamond(\partial B),$$

which will play an important role in Chapter 2 (compare Theorem 2.2.6).

In addition, we will need appropriate function spaces inside the domain B. First of all, we define the space

$$L^2(B, \mathbb{R}^d) := \left\{ h : B \to \mathbb{R}^d : \iint_B |h|^2 \, dx < \infty \right\},$$

which we will mainly need in Chapter 3.

The electrical potential u that solves the direct problem of EIT is usually considered as an element of the Sobolev space $H^1(B)$. A connection to the previously defined Sobolev spaces on the boundary is given by the trace theorem (see e.g. Chapter 3 in [66]) which states that for $u \in H^1(B)$ the trace $u|_{\partial B}$ is an element of $H^{\frac{1}{2}}(\partial B)$ and that the trace operator that maps $u \in H^1(B)$ to $u|_{\partial B}$ is bounded.

Last but not least, we need to define the following closed subspace of $H^1(B)$:

$$H^1_\diamond(B) = \{u \in H^1(B) : u|_{\partial B} \in H^{\frac{1}{2}}_\diamond(\partial B)\},$$

in which we will look for solutions of the direct problem in the following section.

1.2.3 The Direct Problem

Now we have collected all the tools to formulate the direct problem in the subject $B \subset \mathbb{R}^d$ ($d = 2, 3$) which is assumed to be a bounded and simply connected C^2- domain.

The conductivity $\gamma : B \to \mathbb{C}$ is assumed to be such that $\gamma \in L^\infty(B)$. Furthermore, we assume that $\mathrm{Re}\,\gamma$ is bounded from below by a constant $c > 0$ and that $\mathrm{Im}\,\gamma$ is non-positive almost everywhere in B.

For a given current pattern f at ∂B we now want to find a an electrical potential u that solves the Neumann boundary value problem

$$\mathrm{div}(\gamma \nabla u) = 0 \text{ in } B, \quad \gamma \frac{\partial u}{\partial \nu} = f \text{ on } \partial B. \tag{1.3}$$

A weak formulation for the Neumann boundary value problem with boundary values $f \in L^2(\partial B)$ is to find $u \in H^1(B)$ that solves

$$\iint_B \gamma \nabla u \cdot \nabla \overline{\psi} \, dx = \int_{\partial B} f \overline{\psi} \, ds \quad \text{for all } \psi \in H^1(B). \tag{1.4}$$

By setting $\psi \equiv 1$ in B we observe that for existence of a solution $u \in H^1(B)$ to (1.4) the current pattern f has to satisfy $f \in L^2_\diamond(\partial B)$. Since we restrict the Neumann boundary values to $L^2_\diamond(\partial B)$, the problem (1.4) doesn't change if the test functions ψ are restricted to $H^1_\diamond(B)$. In addition,

u can be determined uniquely at most up to an additive constant. We therefore also restrict our solutions u to $H^1_\diamond(B)$.

As a result we obtain the following weak formulation: for a given current pattern $f \in L^2_\diamond(\partial B)$ find $u \in H^1_\diamond(B)$ that solves

$$\iint_B \gamma \nabla u \cdot \nabla \overline{\psi} \, dx = \int_{\partial B} f \overline{\psi} \, ds \quad \text{for all } \psi \in H^1_\diamond(B). \tag{1.5}$$

Now the right hand of (1.5) side leads to a linear form $L : \psi \mapsto \int_{\partial B} f \overline{\psi} \, ds$ on $H^1_\diamond(B)$ for which we can estimate

$$|L\psi| \leq \|f\|_{L^2(\partial B)} \|\psi\|_{L^2(\partial B)} \leq c \|f\|_{L^2(\partial B)} \|\psi\|_{H^1(B)}$$

for some constant $c > 0$, hence L is bounded on $H^1_\diamond(B)$. On the left hand side of (1.5) the bilinear form $a(u, \psi) = \iint_B \gamma \nabla u \cdot \nabla \overline{\psi} \, dx$ on $H^1_\diamond(B)$ is bounded since

$$|a(u, \psi)| \leq \text{ess sup}\{\gamma\} \|\nabla u\|_{L^2(B)} \|\nabla \psi\|_{L^2(B)} \leq \tilde{c} \|u\|_{H^1(B)} \|\psi\|_{H^1(B)}$$

for some constant $\tilde{c} > 0$. Coercivity of a can be shown using the lower bound c for $\text{Re}\,\gamma$ and Poincaré's inequality:

$$\text{Re}\{a(u, u)\} \geq c \|\nabla u\|^2_{L^2(B)} \geq \hat{c} \|u\|^2_{H^1(B)}$$

for some constant $\hat{c} > 0$. Using the Lax-Milgram theorem (see [66] or [21]) it can now be shown that the problem (1.5) is uniquely solvable for every $f \in L^2_\diamond(\partial B)$ and that it is well-posed, i.e. the potential u depends continuously on the Neumann data f. Throughout this work the solution of the direct problem is meant in the sense of (1.5).

The trace theorem yields together with the well-posedness of the direct problem that the Neumann-to-Dirichlet operator

$$\Lambda : L^2_\diamond(\partial B) \to L^2_\diamond(\partial B), \quad f \mapsto u|_{\partial B},$$

where $u \in H^1_\diamond(B)$ solves (1.5) is well-defined and bounded.

Now that we have introduced the weak formulation (1.5) we may also consider the conormal derivative ∂_γ that corresponds to the normal derivative in the classical formulation (1.3). In Chapter 4 of [66] it is shown that for a weak solution $u \in H^1(B)$ of the partial differential equation $\mathrm{div}(\gamma\nabla u) = 0$ the conormal derivative at ∂B is an element of $H^{-\frac{1}{2}}(\partial B)$ and the map $u \mapsto \partial_\gamma u$ is also bounded.

In addition, it can be shown that Λ is an isomorphism from $H_\diamond^{-\frac{1}{2}}(\partial B)$ to $H_\diamond^{\frac{1}{2}}(\partial B)$ (see e.g. [30]). Furthermore, this result also implies that Λ has dense range in $L_\diamond^2(\partial B)$.

1.3 The Inverse Problem

The inverse problem of electrical impedance tomography (or inverse conductivity problem) is to recover the conductivity γ from the knowledge of the Neumann-to-Dirichlet-map Λ. This inverse problem has first been posed by Calderon in the influential paper [15]. Since then, various proofs for uniqueness of the inverse problem, i.e. the unique identifiability of γ from the knowledge of Λ, have been developed for different classes of conductivities and dimensions. We proceed by giving a short outline of identifiability results.

1.3.1 Identifiability and Reconstruction Methods

In [58] Kohn and Vogelius showed unique identifiability of real-valued piecewise analytic conductivities, and Sylvester and Uhlmann proved in [73] uniqueness for real-valued C^2-conductivities in dimensions $d \geq 3$. In Section 5.7 of [47] a uniqueness result for piecewise C^2-conductivities is shown. For the dimension $d = 2$ uniqueness for real-valued $W^{2,p}$ conductivities was shown by Nachman in [68], and Astala and Päivärinta proved uniqueness even for real-valued L^∞ conductivities in [5]. The most general uniqueness result for real-valued conductivities in dimensions $d \geq 3$ is shown by Brown and Torres in [7] so far. A uniqueness result for complex-valued conductivities is shown by Francini in [24].

These identifiability results also led to reconstruction methods out of which a very successful one is the d-bar method (see e.g. [57] or the review paper [75]).

There are some stability results for the determination of the conductivity from the Neumann-to-Dirichlet-map (see e.g. [1]). However, they require rather restrictive regularity assumptions on the conductivity. In

Section 2.3 of [8] and in Chapter 7 of [70] a simple example for radially symmetric conductivities demonstrates the severe ill-posedness of the inverse conductivity problem

There is a large variety of reconstruction methods for EIT out of which we only mention a few. A very successful class of reconstruction methods is based on the linearization using the Fréchet derivative of Λ with respect to γ. This linearization can be used to develop iterative reconstruction methods as well as one-step algorithms (see e.g. [17], [22] and [65]).

Another class of reconstruction methods uses the a priori assumption that there is an inclusion, i.e. a domain inside the investigated subject with a different conductivity than the background medium. The main idea is the linearization of the problem using the Fréchet derivative with respect to the boundary of the inclusion. This Fréchet derivative is called domain derivative, and the linearization leads to iterative reconstruction methods as in [43] or [23].

The assumption that there is an inclusion is also fundamental for the reconstruction approaches in [40], [41] and [42] which are adaptions of the concept of the convex scattering support and the convex backscattering support from [60] and [38] to the inverse problem of EIT.

For a more extensive overview over various aspects of impedance tomography and reconstruction methods we refer to the review articles [16],[6] and [75].

1.3.2 The Factorization Method

This work is also focused in the detection and characterization of anomalies or inclusions, i.e. of domains in which the conductivity is different from the background medium. In order to formulate the corresponding inverse problem we state further a priori assumptions. Let Ω be a C^2-domain such that $\overline{\Omega} \subset B$ and $B \setminus \overline{\Omega}$ is connected. Furthermore, let the conductivity be as follows:

$$\gamma(x) = \gamma_0(x) + \chi_\Omega(x)\gamma_1(x),$$

where χ_Ω denotes the characteristic function of the domain Ω. γ_0 is called the background conductivity and is assumed to be known a priori and sufficiently smooth.

The problem that we consider in Chapter 2 is to recover the inclusion Ω from the Neumann-to-Dirichlet map Λ. More precisely, let Λ_0 : $L^2_\diamond(\partial B) \to L^2_\diamond(\partial B)$ be the Neumann-to-Dirichlet map for the background case, i.e. for the case in which the conductivity is equal to γ_0 everywhere in B. In particular, $\Lambda_0 : L^2_\diamond(\partial B) \to L^2_\diamond(\partial B)$, and $f \mapsto u_0|_{\partial B}$, where $u_0 \in H^1_\diamond(B)$ solves (1.5) for γ_0 instead of γ. Now the problem is to recover the inclusion Ω from the difference $\Lambda - \Lambda_0$.

The method that we use to solve this inverse problem is the Factorization method that has first been suggested by Kirsch in [51] for inverse obstacle scattering problems. Since then the method has been applied to a wide range of problems such as scattering by an inhomogeneous medium (see [52]) or scattering from obstacles with mixed boundary conditions (see [33], [37], [35] and [36]). The method has also been developed for scattering from periodic structures and rough surfaces (see [4], [3], [62]) and for Maxwell's equations (see [54]). The monograph [56] presents many aspects and applications of the method.

The Factorization method for impedance tomography has first been developed by Brühl and Hanke in [8], [10], [9] and [11]. The review article [39] provides an overview over these papers. Since then the method experienced further analysis, extensions and further developments. In [61] a regularization technique for the Factorization method is developed, while in [53] and [25] more general partial differential equations are under consideration. In [32] the detection of infinitely thin inclusions is considered. Furthermore, there are several works dealing with weaker assumptions on the conductivity (see e.g. [45] [27], [28]).

The assumption that Λ is known, i.e. that the whole surface potential $u|_{\partial B}$ is known for every current pattern $f \in L^2_\diamond(\partial B)$ is an idealized model called the continuum model. In real applications the current is applied and the voltage is measured through a finite number of electrodes. The most realistic way to model these electrodes is the complete electrode model (see [18] or [72]). The Factorization method has successfully been applied in the framework of the complete electrode model in [44], [46] and [64]. However, for the sake of simplicity we restrict to the continuum model in this work.

1.3.3 Outline of this Work

This work consists of two main parts and contributes to the Factorization method for EIT as follows.

One drawback of the Factorization method for EIT is that in order to prove it to work one has to assume that either all the inclusions (i.e. all the components of Ω) have a higher or a lower conductivity than the background medium. In Chapter 2 we therefore present a modified version of the Factorization method which we can prove to work even in the mixed case, i.e. in the case of both inclusions with a lower conductivity and inclusions with a higher conductivity than the background medium. This modification is mainly is based on the considerations in [34], [33] and [37] where a modified version of the Factorization method for inverse scattering problems with different types of obstacles is presented.

In Section 2.1 we therefore first investigate the Factorization method for EIT in the case of at least two disjoint inclusions. We show that in the case of mixed inclusions, the usual proof for the Factorization method to work fails. The main reason for this is that in the case of mixed inclusions the operator that appears in the middle of the factorization fails to be coercive.

Afterwards, in Section 2.2 we adapt the method suggested by Grinberg and Kirsch in [34], [33] and [37] for acoustic scattering problems to derive the Factorization method for a slightly different factorization to the inverse problem of detecting inclusions in EIT. The method is mainly based on the idea to cover one type of inclusions with a synthetic inclusion to obtain a factorization in which the middle operator is coercive. As a result we obtain a two-step algorithm in which we first cover one type of inclusions to reconstruct the other type and then vice versa.

In Section 2.3 we present some numerical tests with this new method and compare the results to the results obtained with the original Factorization method which is not proven to work in the mixed case.

Since the Factorization method (or any other qualitative method) provides information about location and shape of inclusions but not about their actual conductivity, it is desirable to have a method to obtain this conductivity after the inclusions have been identified. In Chapter 3 we present such a method that is also closely related to the Factorization method.

This method is based on a new version of the Factorization method for

EIT that involves a factorization with three operators that are different from those in e.g. Chapter 2. In Section 3.1 we present this new version of the method and introduce the operators involved.

Afterwards, in Section 3.2 we investigate the spectrum of the operator in the middle of the factorization for the case of piecewise constant conductivities and show bounds for it. This spectrum is also closely related to the spectrum of boundary integral operators. In addition, we show that the spectrum of the middle operator exhibits one accumulation point for the case of only one inclusion and N different accumulation points in the case of N inclusions with different conductivities. These accumulation points have a direct connection to the desired conductivity contrasts.

Since the information about the conductivity inside anomalies can be obtained from knowledge about the spectrum of the middle operator of the factorization, in Section 3.3 we develop numerical methods to obtain an approximation of the spectrum from a generalized eigenvalue problem that involves only operators that are known after the inclusions have been detected. We also implemented some numerical tests with this new method that are described in Section 3.4.

2 The Factorization Method for EIT in the Case of Mixed Inclusions

For the Factorization method in EIT it is usually assumed that there is only one type of inclusions present, e.g. only inclusions with a higher or with a lower conductivity than the background. To our knowledge there is no proof that the method works for mixed inclusions, i.e. the case in which there are inclusions with a lower as well as inclusions with a higher conductivity than the background medium. This is a severe restriction since there are applications for EIT in which such mixed conductivity distributions do occur, e.g. thorax imaging.

The reason for this restrictive assumption is that the proof of the main result of the method, i.e. the range identity, relies on the fact that the operator that appears in the middle of the factorization is either positively or negatively coercive or at least a compact perturbation of such an operator. For a mixed conductivity distribution this is no longer fulfilled.

There are several recent works dealing with related problems of weakening assumptions on the problem setting for the Factorization method. In the article [27] it is shown that the assumption on the contrast between the background conductivity and the inclusion conductivity can be weakened but it is still assumed that the conductivity inside all inclusions is either higher or lower than in the background. In [28] a slightly different partial differential equation is considered. Again, the contrasts in the coefficients of the differential equation need to have the same direction for all inclusions.

A very related problem occurs for the Factorization method in acoustic scattering if there are both sound-soft and sound-hard obstacles. This problem has been considered by Grinberg and Kirsch in [34], [33], [37] and [56]. They introduced additional a priori information about the location of the obstacles in order to derive the Factorization method for slight modifications of the far field operator. For this modified version of the Factorization method it can be shown that the method works in the case of both sound-soft and sound-hard obstacles.

The aim of this chapter is to apply these techniques to EIT. At first, in Section 2.1 we formulate the Factorization method for EIT according to [53] and illustrate why we cannot prove that the method works in the mixed case. In Section 2.2 we present the modified version of the Factorization method and show that it is capable of detecting inclusions even in the mixed case, while in Section 2.3 we show some numerical examples using the new method.

2.1 The Factorization Method for two Inclusions

In this section we first summarize the main steps for the Factorization method in EIT as it is formulated in [53]. Afterwards we take a closer look at the middle operator in the case of two disjoint inclusions. These considerations serve as a basis for the covering method that is presented in the following section.

2.1.1 The Standard Factorization Method

We start by explaining the problem setting and by providing an overview over the original Factorization method for impedance tomography. Before we can define the operators that appear in the factorization, we need to state some basic assumptions on the underlying geometry of the body considered and the inclusions as well as the conductivity distribution. Since our considerations are aimed at the detection of mixed inclusion types, we adapt our notations to the case of at least two disjoint inclusions.

Assumption 2.1.1. Let $B \subset \mathbb{R}^d$ ($d \in \{2,3\}$) be a bounded C^2-domain. Let Ω_1, Ω_2 be subdomains of B with C^2-boundaries, $\overline{\Omega}_1, \overline{\Omega}_2 \subset B$ and $\overline{\Omega}_1 \cap \overline{\Omega}_2 = \emptyset$. By Ω we denote the union of both inclusions: $\Omega = \Omega_1 \cup \Omega_2$. Let $B \setminus \overline{\Omega}$ be connected.

Additionally, we allow isotropic and complex-valued conductivities that have properties as stated below.

Assumption 2.1.2. Let the conductivity be as follows: $\gamma : B \to \mathbb{C}$ and

$$\gamma(x) = \begin{cases} \gamma_0(x), & x \in B \setminus \overline{\Omega}, \\ \gamma_0(x) + \gamma_1(x), & x \in \Omega_1, \\ \gamma_0(x) + \gamma_2(x), & x \in \Omega_2. \end{cases}$$

Let the background conductivity γ_0 be real-valued, satisfy $\gamma_0 \in C^{2,\alpha}(\overline{B})$ for some $\alpha > 0$ and let let γ_0 be strictly positive, i.e. there exists $c_0 > 0$ such that

$$\gamma_0(x) \geq c_0 > 0 \quad \text{for all } x \in B.$$

For γ we require that $\operatorname{Re} \gamma$ is strictly positive, i.e. there exists $c > 0$ such that

$$\operatorname{Re} \gamma(x) \geq c > 0 \quad \text{for almost all } x \in B.$$

We assume that $\gamma_j \in L^\infty(\Omega_j)$ ($j = 1, 2$) and allow two different types of inclusions: for $j = 1, 2$ there exists $c_j > 0$ such that either

$$\operatorname{Re} \gamma_j(x) \leq -c_j < 0 \quad \text{for almost all } x \in \Omega_j \text{ or}$$

$$\operatorname{Re} \gamma_j(x) \geq c_j > 0 \quad \text{for almost all } x \in \Omega_j.$$

In the first case we call Ω_j an inclusion of type 1 and in the second case an inclusion of type 2.

The inclusions may have a complex-valued conductivity with negative imaginary part. In particular, we assume that for $j = 1, 2$ either $\operatorname{Im} \gamma_j(x) = 0$ holds for almost all $x \in \Omega_j$ or $\operatorname{Im} \gamma_j(x) < 0$ holds for almost all $x \in \Omega_j$ (see also Remark 2.1.6).

Assumption 2.1.2 means in particular that type 1 inclusions have a lower absolute conductivity than the background, while type 2 inclusions have a higher absolute conductivity than the background.

We proceed by deriving the Factorization method according to [53] but for the special case of two disjoint inclusions. We only recall the main results while for the corresponding proofs we refer to [53]. Firstly, we define the operators that appear in the factorization of $\Lambda - \Lambda_0$. Consider the operator

$$G : H_\diamond^{-\frac{1}{2}}(\partial\Omega_1) \times H_\diamond^{-\frac{1}{2}}(\partial\Omega_2) \to L_\diamond^2(\partial B), \quad (g_1, g_2)^\top \mapsto v|_{\partial B},$$

where $v \in H^1(B \setminus \overline{\Omega})$ with $v|_{\partial B} \in H_\diamond^{\frac{1}{2}}(\partial B)$ solves

$$\iint_{B \setminus \Omega} \gamma_0 \nabla v \cdot \nabla \overline{\psi} \, dx = \int_{\partial \Omega_1} g_1 \overline{\psi} \, ds + \int_{\partial \Omega_2} g_2 \overline{\psi} \, ds \quad \text{for all } \psi \in H^1(B \setminus \overline{\Omega}).$$

(2.1)

This weak formulation corresponds to the following Neumann boundary value problem: find $v \in H^1(B \setminus \overline{\Omega})$ such that $v|_{\partial B} \in H_\diamond^{\frac{1}{2}}(\partial B)$ and

$$\begin{aligned}
\operatorname{div}(\gamma_0 \nabla v) &= 0 \text{ in } B \setminus \overline{\Omega}, \\
\partial_{\gamma_0} v &= g_j \text{ on } \partial \Omega_j \ (j = 1, 2), \\
\partial_{\gamma_0} v &= 0 \text{ on } \partial B.
\end{aligned}$$

As shown in [53], G is compact and one-to-one, $\mathcal{R}(G)$ is dense in $L_\diamond^2(\partial B)$, and the corresponding assertions hold for its adjoint $G^* : L_\diamond^2(\partial B) \to H_\diamond^{\frac{1}{2}}(\partial \Omega_1) \times H_\diamond^{\frac{1}{2}}(\partial \Omega_2)$.

Furthermore, the following operator will appear:

$$T : H_\diamond^{\frac{1}{2}}(\partial \Omega_1) \times H_\diamond^{\frac{1}{2}}(\partial \Omega_2) \to H_\diamond^{-\frac{1}{2}}(\partial \Omega_1) \times H_\diamond^{-\frac{1}{2}}(\partial \Omega_2),$$

$$(h_1, h_2)^\top \mapsto \left(\partial_{\gamma_0} w|_{+,1}, \partial_{\gamma_0} w|_{+,2} \right)^\top,$$

where $w \in H^1(B \setminus \partial \Omega)$ with $w|_{\partial B} \in H_\diamond^{\frac{1}{2}}(\partial B)$ satisfies the jump conditions $w|_{+,j} - w|_{-,j} = h_j \ (j = 1, 2)$ and

$$\iint_B \gamma \nabla w \cdot \nabla \overline{\psi} \, dx = 0 \quad \text{for all } \psi \in H_\diamond^1(B). \tag{2.2}$$

The notation $w|_\pm$ denotes the trace of w from the exterior and interior of Ω_j, respectively. In addition, we use the notation $w|_{\pm,j}$ for the traces from exterior and interior at the particular boundary $\partial \Omega_j \ (j = 1, 2)$. The corresponding classical formulation is to find $w \in H^1(B \setminus \partial \Omega)$ such that

$w|_{\partial B} \in H^{\frac{1}{2}}_{\diamond}(\partial B)$ that solves the transmission problem

$$\begin{aligned}
\operatorname{div}(\gamma \nabla w) &= 0 \text{ in } B \setminus \partial \Omega, \\
\partial_{\gamma_0} w &= 0 \text{ on } \partial B, \\
\partial_{\gamma_0} w|_+ - \partial_\gamma w|_- &= 0 \text{ on } \partial \Omega_j \ (j = 1, 2), \\
w|_+ - w|_- &= h_j \text{ on } \partial \Omega_j \ (j = 1, 2).
\end{aligned} \tag{2.3}$$

The operator $T_0 : H^{\frac{1}{2}}_{\diamond}(\partial \Omega_1) \times H^{\frac{1}{2}}_{\diamond}(\partial \Omega_2) \rightarrow H^{-\frac{1}{2}}_{\diamond}(\partial \Omega_1) \times H^{-\frac{1}{2}}_{\diamond}(\partial \Omega_2)$ is defined just the same as T but with γ replaced by γ_0 in (2.2), i.e. T_0 corresponds to T in the background case.

In order to show well-definedness and boundedness of T we derive a weak formulation equivalent to the problem (2.2) and the additional jump conditions at $\partial \Omega_j$ $(j = 1, 2)$.

For $h_j \in H^{\frac{1}{2}}_{\diamond}(\partial \Omega_j)$ choose $\hat{w}^{(j)} \in H^1(\Omega_j)$ such that $\hat{w}^{(j)}\big|_{\partial \Omega_j} = h_j$ $(j = 1, 2)$. Then by setting $\tilde{w} = w + \chi_{\Omega_1}\hat{w}^{(1)} + \chi_{\Omega_2}\hat{w}^{(2)}$ we obtain the formulation: find $\tilde{w} \in H^1_{\diamond}(B)$ satisfying

$$\iint_B \gamma \nabla \tilde{w} \cdot \nabla \overline{\psi} \, dx = \iint_{\Omega_1} \gamma \nabla \hat{w}^{(1)} \cdot \nabla \overline{\psi} \, dx + \iint_{\Omega_2} \gamma \nabla \hat{w}^{(2)} \cdot \nabla \overline{\psi} \, dx \tag{2.4}$$

for all $\psi \in H^1_{\diamond}(B)$. Since $w = \tilde{w}$ in $B \setminus \overline{\Omega}$ the definition of T is equivalent to setting $(h_1, h_2)^\top \mapsto \left(\partial_{\gamma_0}\tilde{w}|_{+,1}, \partial_{\gamma_0}\tilde{w}|_{+,2} \right)^\top$ and $\tilde{w}$ solves (2.4). Here we can also show unique solvability using the Lax Milgram theorem, and from the trace theorem we obtain well-definedness and boundedness of T, T_0.

Now we state the well-known factorization that can be proven independently of the signs of γ_1 and γ_2,

$$\Lambda - \Lambda_0 = G(T - T_0)G^*. \tag{2.5}$$

Figure 2.1 illustrates this factorization, and for its proof we refer to [53].

The adjoint T^* of T is characterized just as T but in the transmission boundary value problem (2.3) (and thus also in (2.4)) γ is replaced by $\overline{\gamma}$. This implies for real-valued conductivities that T is self-adjoint and in particular that T_0 is self-adjoint.

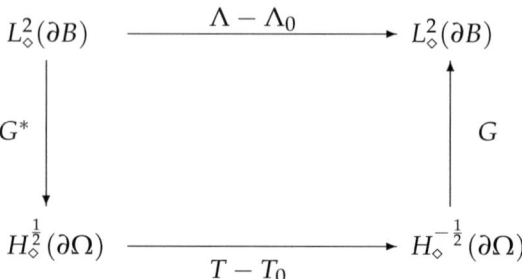

Figure 2.1: Diagram of operators and spaces in the factorization (2.5)

Moreover, we can define the operators $\mathrm{Re}\, T := \frac{1}{2}\left(T + T^*\right)$ and $\mathrm{Im}\, T := \frac{1}{2i}(T - T^*)$, and the equality $T = \mathrm{Re}\, T + i\,\mathrm{Im}\, T$ holds obviously. Now for $\mathrm{Im}\, T$ it is shown in [53] that

$$\langle(\mathrm{Im}\, T)h, h\rangle = \mathrm{Im}\,\langle Th, h\rangle = -\int_{\Omega} \mathrm{Im}\,\gamma\,|\nabla w|^2 \geq 0 \quad \text{for all } h \in H_\diamond^{\frac{1}{2}}(\partial\Omega). \tag{2.6}$$

Under the assumption that only one inclusion type is present inside B the real part of the middle operator $T - T_0$ satisfies the following coercivity assertions (Theorem 2.3 in [53]):

Lemma 2.1.3. *Let T, T_0 be defined as above.*

(a) If Ω_1 and Ω_2 are both of type 1, then there is $c > 0$ such that

$$\langle(\mathrm{Re}\, T - T_0)h, h\rangle \geq c\,\|h\|^2_{H^{\frac{1}{2}}(\partial\Omega)}$$

for all $h \in H_\diamond^{\frac{1}{2}}(\partial\Omega)$,

(b) If Ω_1 and Ω_2 are both of type 2, then there is $c > 0$ such that

$$\langle-(\mathrm{Re}\, T - T_0)h, h\rangle \geq c\,\|h\|^2_{H^{\frac{1}{2}}(\partial\Omega)}$$

for all $h \in H_\diamond^{\frac{1}{2}}(\partial\Omega)$.

In the following sections the case of mixed inclusions will be considered. However, we will show connection to the cases in which there is only one of the two inclusion types present. We therefore define the following modified versions of T for $j = 1, 2$: $T^{(j)} : H_\diamond^{\frac{1}{2}}(\partial\Omega_j) \to H_\diamond^{-\frac{1}{2}}(\partial\Omega_j)$ and it maps $h_j \mapsto \partial_{\gamma_0}\tilde{w}^{(j)}\big|_{j,+}$. As for (2.4) we choose $\hat{w}^{(j)} \in H^1(\Omega_j)$ such that $\hat{w}^{(j)}\big|_{\partial\Omega_j} = h_j$ and $\tilde{w}^{(j)} \in H^1(B \setminus \partial\Omega_j)$ with $\tilde{w}^{(j)}\big|_{\partial B} \in H_\diamond^{\frac{1}{2}}(\partial B)$ solves

$$\iint_B \gamma_0 \nabla\tilde{w}^{(j)} \cdot \nabla\overline{\psi}\,dx + \iint_{\Omega_j} \gamma_j \nabla\tilde{w}^{(j)} \cdot \nabla\overline{\psi}\,dx = \iint_{\Omega_j} \gamma\nabla\hat{w}^{(j)} \cdot \nabla\overline{\psi}\,dx \quad (2.7)$$

for all $\psi \in H^1(B)$.

Now we observe that $T^{(j)}$ is equal to T for the case in which Ω_j is the only inclusion inside B. In addition, we define $T_0^{(j)} : H_\diamond^{\frac{1}{2}}(\partial\Omega_j) \to H_\diamond^{-\frac{1}{2}}(\partial\Omega_j)$ just the same but with $\gamma \equiv \gamma_0$ everywhere in B. Now Lemma 2.1.3 implies the following assertion.

Lemma 2.1.4. *Let $T^{(j)}$, $T_0^{(j)}$ be defined as above for $j = 1, 2$, then there is $c > 0$ such that:*

(a) $\left\langle (\mathrm{Re}\, T^{(1)} - T_0^{(1)})h, h \right\rangle \geq c\,\|h_1\|_{H^{\frac{1}{2}}(\partial\Omega)}^2$ *for all $h_1 \in H_\diamond^{\frac{1}{2}}(\partial\Omega_1)$,*

(b) $\left\langle (\mathrm{Re}\, T_0^{(2)} - T^{(2)})h, h \right\rangle \geq c\,\|h_2\|_{H^{\frac{1}{2}}(\partial\Omega)}^2$ *for all $h_2 \in H_\diamond^{\frac{1}{2}}(\partial\Omega_2)$.*

Another immediate conclusion from Lemma 2.1.3 is that the operator $\mathrm{Re}\, T - T_0$ is one-to-one. Later we will see that these coercivity results are not valid in the mixed case. However, we are still able to show the following Lemma, which we will need in Section 2.2.

Lemma 2.1.5. *Assume that Ω_1 is of type 1 while Ω_2 is of type 2, then:*

(a) *The operator $T - T_0$ is one-to-one.*

(b) *For $0 \neq h = (h_1, h_2)^\top \in H_\diamond^{\frac{1}{2}}(\partial\Omega_1) \times H_\diamond^{\frac{1}{2}}(\partial\Omega_2)$ with $(\mathrm{Re}\, T - T_0)h = 0$ it holds that $\mathrm{Im}\,\langle Th, h \rangle > 0$.*

Proof. Part a): Let $h = (h_1, h_2)^\top \in H_\diamond^{\frac{1}{2}}(\partial\Omega_1) \times H_\diamond^{\frac{1}{2}}(\partial\Omega_2)$ be such that $(T - T_0)h = 0$. Then

$$\partial_{\gamma_0} w|_{+,1} = \partial_{\gamma_0} w_0|_{+,1}, \quad \partial_{\gamma_0} w|_{+,2} = \partial_{\gamma_0} w_0|_{+,2},$$

where w, w_0 are weak solutions to the transmission boundary value problem (2.3) for the actual conductivity and the background case, respectively. By uniqueness of the Neumann problem in $B \setminus \overline{\Omega}$ we know that $w \equiv w_0$ in $B \setminus \overline{\Omega}$, from which we can conclude, using the jump conditions at $\partial\Omega_1, \partial\Omega_2$, that

$$\partial_\gamma w|_{-,1} = \partial_{\gamma_0} w_0|_{-,1}, \quad \partial_\gamma w|_{-,2} = \partial_{\gamma_0} w_0|_{-,2},$$

$$w|_{-,1} = w_0|_{-,1}, \quad w|_{-,2} = w_0|_{-,2}.$$

Using the first Green identity it yields

$$\iint\limits_{\Omega_j} \gamma |\nabla w|^2 \, dx = \iint\limits_{\Omega_j} \gamma_0 |\nabla w_0|^2 \, dx \quad (j = 1, 2). \tag{2.8}$$

We have to distinguish between two different cases. If $\mathrm{Im}\,\gamma < 0$ in Ω_1, then we obtain $\nabla w \equiv \nabla w_0 \equiv 0$ in Ω_1 immediately. If γ is real-valued inside Ω_1 then we have to argue as follows: we use the first Green identity to show

$$\iint\limits_{\Omega_1} \gamma \nabla w \cdot \nabla(\overline{w} - \overline{w}_0) \, dx = \int\limits_{\partial\Omega_1} (\overline{w} - \overline{w}_0) \partial_\gamma w \, ds = 0, \tag{2.9}$$

and estimate using (2.9)

$$0 \leq \iint\limits_{\Omega_1} \gamma |\nabla(w - w_0)|^2 \, dx$$

$$= \iint\limits_{\Omega_1} \gamma \nabla w \cdot \nabla(\overline{w} - \overline{w}_0) \, dx - \iint\limits_{\Omega_1} \gamma \nabla w_0 \cdot \nabla(\overline{w} - \overline{w}_0) \, dx$$

$$= \iint\limits_{\Omega_1} \gamma |\nabla w_0|^2 \, dx - \iint\limits_{\Omega_1} \gamma \nabla w_0 \cdot \nabla \overline{w} \, dx.$$

Using (2.9) once more yields

$$\iint\limits_{\Omega_1} \gamma |\nabla w_0|^2 \, dx \geq \iint\limits_{\Omega_1} \gamma |\nabla w|^2 \, dx.$$

This inequality and the fact that Ω_1 is a type 1 inclusion allows us to estimate

$$\iint\limits_{\Omega_1} \gamma |\nabla w|^2 \, dx = \iint\limits_{\Omega_1} \gamma_0 |\nabla w_0|^2 \, dx$$

$$\geq \iint\limits_{\Omega_1} \gamma |\nabla w_0|^2 \, dx + c_1 \iint\limits_{\Omega_1} |\nabla w_0|^2 \, dx$$

$$\geq \iint\limits_{\Omega_1} \gamma |\nabla w|^2 \, dx + c_1 \iint\limits_{\Omega_1} |\nabla w_0|^2 \, dx.$$

Now it follows that $\nabla w_0 \equiv 0$ and from (2.8) that $\nabla w \equiv 0$ in Ω_1. Hence w and w_0 are constant inside Ω_1. We have to apply the very same arguments to the inclusion Ω_2 using its type 2 property to show that also $\nabla w \equiv \nabla w_0 \equiv 0$ in Ω_2. This means that the normal derivatives of w, w_0 at $\partial \Omega_1$ and at $\partial \Omega_2$ from both sides are zero, which implies that $w = w_0 = 0$ in $B \setminus \overline{\Omega}$. Now h_1 and h_2 have to be constant, and since $h_j \in H_\diamond^{\frac{1}{2}}(\partial \Omega_j)$ $(j = 1, 2)$ we obtain $h_1 = 0$ and $h_2 = 0$.

Part b): Case 1: If $\operatorname{Im} \gamma = 0$ almost everywhere in Ω_1 and in Ω_2, then $\operatorname{Re} T - T_0 = T - T_0$ which is injective according to part a).

Case 2: Now we assume that both $\operatorname{Im} \gamma_1 < 0$ and $\operatorname{Im} \gamma_2 < 0$ almost everywhere in Ω_1 and Ω_2, respectively. From (2.6) we obtain

$$\operatorname{Im} \langle Th, h \rangle = - \iint\limits_{\Omega} \operatorname{Im} \gamma |\nabla w|^2 \, dx.$$

Hence $\operatorname{Im} \langle Th, h \rangle = 0$ implies that $w = const$ in Ω_1 and in Ω_2, from which we deduce that w solves the homogeneous Neumann problem in $B \setminus \overline{\Omega}$ and therefore $w \equiv 0$ in $B \setminus \overline{\Omega}$. Since $h \in H_\diamond^{\frac{1}{2}}(\partial \Omega_1) \times H_\diamond^{\frac{1}{2}}(\partial \Omega_2)$ it follows as in part a) that $h = 0$ and thus that $\operatorname{Im} T$ is positive on $H_\diamond^{\frac{1}{2}}(\partial \Omega_1) \times H_\diamond^{\frac{1}{2}}(\partial \Omega_2)$.

Case 3: The remaining case is the one in which $\operatorname{Im}\gamma < 0$ inside one inclusion and $\operatorname{Im}\gamma = 0$ inside the other inclusion. We assume without loss of generality that $\operatorname{Im}\gamma_1 < 0$ almost everywhere in Ω_1 and $\operatorname{Im}\gamma_2 = 0$ almost everywhere in Ω_2. As before we know that

$$\operatorname{Im}\langle Th, h\rangle = -\iint\limits_{\Omega_1} \operatorname{Im}\gamma_1\,|\nabla w|^2\,dx,$$

and $\operatorname{Im}\langle Th, h\rangle = 0$ implies that $w = const$ in Ω_1. It is easy to show that $\operatorname{Im}\langle Th, h\rangle = -\operatorname{Im}\langle T^*h, h\rangle$ and

$$\operatorname{Im}\langle T^*h, h\rangle = \iint\limits_{\Omega_1} \operatorname{Im}\gamma_1\,|\nabla\tilde{w}|^2\,dx,$$

where $\tilde{w}$ is the weak solution of (2.3) for $\overline{\gamma}$ instead of γ. We deduce that $\tilde{w}$ also has to be constant in Ω_1, and $\partial_{\gamma_0}w|_{+,1} = \partial_{\gamma_0}\tilde{w}|_{+,1} = 0$. Since γ is real-valued outside Ω_1, both w and $\tilde{w}$ are weak solutions to the following transmission boundary value problem: find $w \in H^1(B \setminus (\overline{\Omega}_1 \cup \partial\Omega_2))$ such that $w|_{\partial B} \in H^{\frac{1}{2}}_\diamond(\partial B)$ solves

$$\begin{aligned}
\operatorname{div}(\gamma\nabla w) &= 0 \text{ in } B \setminus (\overline{\Omega}_1 \cup \partial\Omega_2), \\
\partial_{\gamma_0}w &= 0 \text{ on } \partial B \cup \partial\Omega_1, \\
\partial_{\gamma_0}w|_+ - \partial_\gamma w|_- &= 0 \text{ on } \partial\Omega_2, \\
w|_+ - w|_- &= h_2 \text{ on } \partial\Omega_2
\end{aligned}$$

in the weak sense. This problem has a unique solution, thus $\partial_{\gamma_0}w|_{+,2} = \partial_{\gamma_0}\tilde{w}|_{+,2}$, and

$$\operatorname{Im} Th = \frac{1}{2i}\left(\partial_{\gamma_0}w|_{+,1} - \partial_{\gamma_0}\tilde{w}|_{+,1},\, \partial_{\gamma_0}w|_{+,2} - \partial_{\gamma_0}\tilde{w}|_{+,2}\right)^\top = (0,0)^\top.$$

Now let $h \in H^{\frac{1}{2}}_\diamond(\partial\Omega_1) \times H^{\frac{1}{2}}_\diamond(\partial\Omega_2) \neq (0,0)^\top$ be such that $(\operatorname{Re} T - T_0)h = 0$. Then the identity $(T - T_0)h = (\operatorname{Re} T - T_0)h + i\,(\operatorname{Im} T)h = i\,(\operatorname{Im} T)h$ and part a) yield the assertion. □

Remark 2.1.6. Our assumptions on $\operatorname{Im}\gamma$ are rather strict since inclusions in which $\operatorname{Im}\gamma$ is negative only inside a subdomain of Ω_j ($j = 1, 2$) are not allowed. However, under certain smoothness assumptions on γ such as e.g. $\gamma \in C^1(\Omega_j)$ $(j = 1, 2)$ the proof of part b) of Lemma 2.1.5 can be extended to more general conductivities using the unique continuation principle.

Now we return to our outline of the Factorization method and assume that there is only one of the two inclusion types present. The coercivity results from Lemma 2.1.3 play a crucial role in the proof of the range identity

$$\mathcal{R}\left(|\text{Re}\,\Lambda - \Lambda_0|^{\frac{1}{2}}\right) = \mathcal{R}(G), \tag{2.10}$$

since it relies on existence and bijectivity of the operator $|\text{Re}\,T - T_0|^{\frac{1}{2}}$: $H^{\frac{1}{2}}_\diamond(\partial\Omega) \to L^2_\diamond(\partial\Omega)$ (compare Section 2.3 in [53]). This range identity is the main result for the Factorization method, and it leads to a binary criterion that characterizes the inclusion Ω. In order to derive this binary criterion we first have to define dipole potentials.

The Neumann function N for the domain B and the background conductivity γ_0 is for a fixed point $y \in B$ defined as the distributional solution to

$$\text{div}_x\left(\gamma_0 \nabla_x N(x, y)\right) = \delta(x - y) \quad \text{in } B,$$
$$\partial_{\gamma_0, x} N(x, y) = |\partial B|^{-1} \quad \text{on } \partial B, \tag{2.11}$$
$$\int_{\partial B} N(x, y) ds(x) = 0.$$

N is also symmetric, i.e. $N(x, y) = N(y, x)$ for all $x, y \in B$, $x \neq y$ (see e.g. [67]). Using the Neumann function we construct the dipole potential φ_y for every point $y \in B$ by setting

$$\varphi_y(x) = \gamma_0(y)\left(\hat{a} \cdot \nabla_y N(x, y)\right), \tag{2.12}$$

where $\hat{a}$ is an arbitrary but fixed unit vector.

Now the range identity (2.10) can be used in the following way: it can be shown that

$$y \in \Omega_1 \cup \Omega_2 \Leftrightarrow \varphi_y \in \mathcal{R}(G), \tag{2.13}$$

and the proof is mainly based on the singularity of φ_y in y (compare Theorem 2.6 in [53]). Together with the range identity (2.10) and the Picard criterion we obtain the following equivalence: let $y \in B$, then

$$y \in \Omega_1 \cup \Omega_2 \Leftrightarrow \sum_{k=1}^{\infty} \frac{|\langle \varphi_y, \psi_k \rangle|^2}{\lambda_k} < \infty$$

with an eigensystem $\{\lambda_k, \psi_k : k \in \mathbb{N}\}$ of $|\operatorname{Re} \Lambda - \Lambda_0|$.

This binary criterion can now be used to decide whether a point $y \in B$ lies inside the inclusion $\Omega = \Omega_1 \cup \Omega_2$ or not. However, we have to assume that there is only one inclusion type present, which is a severe restriction.

2.1.2 Representations of the Middle Operator

In order to investigate the Factorization method in the special case in which at least two disjoint inclusions are present we now take a closer look at the middle operator $T - T_0$. Our aim is to use the superposition principle to write $T - T_0$ in the form

$$T - T_0 = \begin{pmatrix} T^{(11)} - T_0^{(11)} & T^{(12)} - T_0^{(12)} \\ T^{(21)} - T_0^{(21)} & T^{(22)} - T_0^{(22)} \end{pmatrix}$$

where $T^{(ij)}, T_0^{(ij)} : H_\diamond^{\frac{1}{2}}(\partial\Omega_j) \to H_\diamond^{-\frac{1}{2}}(\partial\Omega_i)$.

In particular, for $h = (h_1, h_2)^\top \in H_\diamond^{\frac{1}{2}}(\partial\Omega_1) \times H_\diamond^{\frac{1}{2}}(\partial\Omega_2)$ these partial operators satisfy

$$\left(T^{(11)} - T_0^{(11)} \right) h_1 + \left(T^{(12)} - T_0^{(12)} \right) h_2 = ((T - T_0)h)|_{\partial\Omega_1},$$

and

$$\left(T^{(21)} - T_0^{(21)} \right) h_1 + \left(T^{(22)} - T_0^{(22)} \right) h_2 = ((T - T_0)h)|_{\partial\Omega_2}.$$

The mapping properties of the partial operators are also illustrated in Figure 2.2.

Furthermore, we derive a connection to the cases in which only one of the two inclusions is present. This will enable us to make use of the coercivity results from Lemma 2.1.4. The subsequent considerations are based on the weak formulation (2.4).

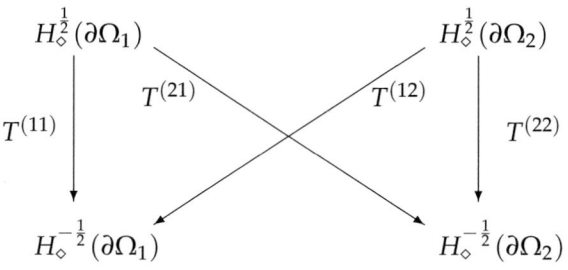

Figure 2.2: Diagram of the partial operators of $T - T_0$ and corresponding spaces

We start by deriving such a representation for T_0. Let $(h_1, h_2)^\top \in H_\diamond^{\frac{1}{2}}(\partial\Omega_1) \times H_\diamond^{\frac{1}{2}}(\partial\Omega_2)$ be given and choose $\hat{w}^{(j)} \in H^1(\Omega_j)$ that satisfies $\hat{w}^{(j)}\big|_{\partial\Omega_j} = h_j$ ($j = 1, 2$). Now consider the following problems: for ($j = 1, 2$) find $\tilde{w}_0^{(jj)} \in H_\diamond^1(B)$ such that

$$\iint_B \gamma_0 \nabla \tilde{w}_0^{(11)} \cdot \nabla \overline{\psi} \, dx = \iint_{\Omega_1} \gamma_0 \nabla \hat{w}^{(1)} \cdot \nabla \overline{\psi} \, dx \quad \text{for all } \psi \in H_\diamond^1(B), \quad (2.14)$$

$$\iint_B \gamma_0 \nabla \tilde{w}_0^{(22)} \cdot \nabla \overline{\psi} \, dx = \iint_{\Omega_2} \gamma_0 \nabla \hat{w}^{(2)} \cdot \nabla \overline{\psi} \, dx \quad \text{for all } \psi \in H_\diamond^1(B). \quad (2.15)$$

By addition of (2.14) und (2.15) we realize that $\tilde{w}_0 := \tilde{w}_0^{(11)} + \tilde{w}_0^{(22)}$ is the solution to (2.4) for the background case, i.e. $\gamma \equiv \gamma_0$. Now we define the components of T_0:

$$T_0^{(11)} : H_\diamond^{\frac{1}{2}}(\partial\Omega_1) \to H_\diamond^{-\frac{1}{2}}(\partial\Omega_1), \quad h_1 \mapsto \partial_{\gamma_0}\tilde{w}_0^{(11)}\big|_{+,1}, \quad \tilde{w}_0^{(11)} \text{ solves (2.14)},$$

$$T_0^{(12)} : H_\diamond^{\frac{1}{2}}(\partial\Omega_2) \to H_\diamond^{-\frac{1}{2}}(\partial\Omega_1), \quad h_2 \mapsto \partial_{\gamma_0}\tilde{w}_0^{(22)}\big|_{+,1}, \quad \tilde{w}_0^{(22)} \text{ solves (2.15)},$$

$$T_0^{(21)} : H_\diamond^{\frac{1}{2}}(\partial\Omega_1) \to H_\diamond^{-\frac{1}{2}}(\partial\Omega_2), \quad h_1 \mapsto \partial_{\gamma_0}\tilde{w}_0^{(11)}\big|_{+,2}, \quad \tilde{w}_0^{(11)} \text{ solves (2.14)},$$

$$T_0^{(22)} : H_\diamond^{\frac{1}{2}}(\partial\Omega_2) \to H_\diamond^{-\frac{1}{2}}(\partial\Omega_2), \quad h_2 \mapsto \partial_{\gamma_0}\tilde{w}_0^{(22)}\big|_{+,2}, \quad \tilde{w}_0^{(22)} \text{ solves (2.15)}.$$

Now we observe that $T_0^{(jj)} = T_0^{(j)}$ holds for $j = 1, 2$, where the operators $T_0^{(j)}$ ($j = 1, 2$) have been defined in the previous section using the variational problem (2.7) in the background case. This means in particular that $T_0^{(11)}$ corresponds to T_0 in the case in which only the inclusion Ω_1 is present, and $T_0^{(22)}$ corresponds to T_0 in the case in which Ω_2 is the only inclusion, respectively. The following lemma provides the desired representation for T_0.

Lemma 2.1.7. $T_0^{(12)}$ *and* $T_0^{(21)}$ *are compact operators. Hence* T_0 *can be represented by*

$$T_0 = \begin{pmatrix} T_0^{(1)} & 0 \\ 0 & T_0^{(2)} \end{pmatrix} + K_0$$

where K_0 *is a compact operator.*

Proof. We only prove the assertion for $T_0^{(12)}$; for $T_0^{(21)}$ the arguments are analogous. Let $\left(h^{(2,j)} \right)_{j \in \mathbb{N}}$ be a bounded sequence in $H_\diamond^{\frac{1}{2}}(\partial \Omega_2)$. Then the corresponding sequence of solutions $\left(\tilde{w}_0^{(22,j)} \right)_{j \in \mathbb{N}}$ of (2.15) is also bounded in $H^1(B)$. Now let $U \subset B \setminus \overline{\Omega}_2$ be a subdomain with $\overline{U} \subset B \setminus \overline{\Omega}_2$ and $\partial \Omega_1 \subset U$. Following Theorem 8.8 in [29] the sequence $\left(\tilde{w}_0^{(22,j)} \right)_{j \in \mathbb{N}}$ is even bounded in $H^2(U)$ and the trace theorem yields that there is a constant c such that $\left\| \partial_{\gamma_0} \tilde{w}_0^{(22,j)} \right\|_{H^{\frac{1}{2}}(\partial \Omega_1)} \leq c$ for all $j \in \mathbb{N}$. From compactness of the imbedding $J : H_\diamond^{\frac{1}{2}}(\partial \Omega_1) \rightarrow H_\diamond^{-\frac{1}{2}}(\partial \Omega_1)$ we obtain the assertion. $\square$

For T the partitioning is not as simple as for T_0, since the conductivity contrasts γ_1, γ_2 appear in Equation (2.4). However, we start as before and divide T into parts contributed by the jump conditions at the individual inclusion boundaries. As before, for $(h_1, h_2)^\top \in H_\diamond^{\frac{1}{2}}(\partial \Omega_1) \times H_\diamond^{\frac{1}{2}}(\partial \Omega_2)$ choose $\hat{w}^{(j)} \in H^1(\Omega_j)$ with $\hat{w}^{(j)} \big|_{\partial \Omega_j} = h_j$ ($j = 1, 2$) and consider the following problems: find $\tilde{w}^{(jj)} \in H^1_\diamond(B)$ $(j = 1, 2)$ satisfying

$$\iint_B \gamma \nabla \tilde{w}^{(11)} \cdot \nabla \overline{\psi} \, dx = \iint_{\Omega_1} \gamma \nabla \hat{w}^{(1)} \cdot \nabla \overline{\psi} \, dx \quad \text{for all } \psi \in H^1_\diamond(B), \quad (2.16)$$

$$\iint_B \gamma \nabla \tilde{w}^{(22)} \cdot \nabla \overline{\psi} \, dx = \iint_{\Omega_2} \gamma \nabla \hat{w}^{(2)} \cdot \nabla \overline{\psi} \, dx \quad \text{for all } \psi \in H^1_\diamond(B). \quad (2.17)$$

One can easily see that $\tilde{w} := \tilde{w}^{(11)} + \tilde{w}^{(22)}$ is a solution to (2.4). As before, we define component operators:

$$T^{(11)} : H^{\frac{1}{2}}_\diamond(\partial\Omega_1) \to H^{-\frac{1}{2}}_\diamond(\partial\Omega_1), \quad h_1 \mapsto \partial_{\gamma_0} \tilde{w}^{(11)}\Big|_{+,1}, \quad \tilde{w}^{(11)} \text{ solves (2.16)},$$

$$T^{(12)} : H^{\frac{1}{2}}_\diamond(\partial\Omega_2) \to H^{-\frac{1}{2}}_\diamond(\partial\Omega_1), \quad h_2 \mapsto \partial_{\gamma_0} \tilde{w}^{(22)}\Big|_{+,1}, \quad \tilde{w}^{(22)} \text{ solves (2.17)},$$

$$T^{(21)} : H^{\frac{1}{2}}_\diamond(\partial\Omega_1) \to H^{-\frac{1}{2}}_\diamond(\partial\Omega_2), \quad h_1 \mapsto \partial_{\gamma_0} \tilde{w}^{(11)}\Big|_{+,2}, \quad \tilde{w}^{(11)} \text{ solves (2.16)},$$

$$T^{(22)} : H^{\frac{1}{2}}_\diamond(\partial\Omega_2) \to H^{-\frac{1}{2}}_\diamond(\partial\Omega_2), \quad h_2 \mapsto \partial_{\gamma_0} \tilde{w}^{(22)}\Big|_{+,2}, \quad \tilde{w}^{(22)} \text{ solves (2.17)}.$$

All four partial operators are well-defined and bounded. However, $T^{(11)}$ is not identical to T in the case in which Ω_1 is the only inclusion. The same holds for $T^{(22)}$. We therefore carry out one more decomposition of these two operators and consider the operators $T^{(1)}$, $T^{(2)}$ defined in (2.7). We continue by investigating $T^{(12)}$ and $T^{(21)}$ as well as the difference between $T^{(11)}$ and $T^{(1)}$ (between $T^{(22)}$ and $T^{(2)}$, respectively).

Lemma 2.1.8. *Let $T^{(ij)}$ and $T^{(i)}$ $(i, j = 1, 2)$ be defined as above. Then:*

(a) *The operators $T^{(12)}$ and $T^{(21)}$ are compact.*

(b) *The operators $S^{(1)} := T^{(11)} - T^{(1)}$ and $S^{(2)} := T^{(22)} - T^{(2)}$ are compact.*

Proof. Part a): We decompose the operator $T^{(12)}$ into two bounded operators out of which one is compact. The proof for $T^{(21)}$ is completely analogous. Therefore let $\hat{\Omega}_1$ be a simply connected C^2-domain in B such that $\overline{\hat{\Omega}_1} \subset B$, $\overline{\Omega}_1 \subset \hat{\Omega}_1$ and $\overline{\hat{\Omega}_1} \cap \overline{\Omega}_2 = \emptyset$. Then $T^{(12)} = \tilde{T} \circ \hat{T}$, where $\hat{T} :$

$H^{\frac{1}{2}}_\diamond(\partial\Omega_2) \rightarrow H^{-\frac{1}{2}}_\diamond(\partial\hat{\Omega}_1)$ maps h_2 to $\partial_{\gamma_0}\tilde{w}^{(22)}\big|_{\partial\hat{\Omega}_1}$, where $\tilde{w}^{(22)} \in H^1_\diamond(B)$

solves (2.17). The second operator $\tilde{T} : H^{-\frac{1}{2}}_\diamond(\partial\hat{\Omega}_1) \rightarrow H^{-\frac{1}{2}}_\diamond(\partial\Omega_1)$ maps g to $\partial_{\gamma_0}v\big|_{1,+}$, where $v \in H^1_\diamond(\hat{\Omega}_1)$ solves

$$\mathrm{div}(\gamma\nabla v) = 0 \text{ in } \hat{\Omega}_1,$$
$$\partial_{\gamma_0}v = g \text{ on } \hat{\Omega}_1$$

in the weak sense. Both $\tilde{T}$ and $\hat{T}$ are bounded operators. Following the same arguments as in the proof of Lemma 2.1.7 it can be shown that $\mathcal{R}(\hat{T}) \subset H^{\frac{1}{2}}_\diamond(\partial\hat{\Omega}_1)$ and thus that $\hat{T}$ is a compact operator.

Part b): The operators $S^{(1)}$ and $S^{(2)}$ are as follows: $S^{(1)} : H^{\frac{1}{2}}_\diamond(\partial\Omega_1) \rightarrow H^{-\frac{1}{2}}_\diamond(\partial\Omega_1)$, $h_1 \mapsto \partial_{\gamma_0}\tilde{v}^{(1)}\big|_{+,1}$ where $\tilde{v}^{(1)} \in H^1_\diamond(B)$ solves

$$\iint_B \gamma\nabla\tilde{v}^{(1)} \cdot \nabla\overline{\psi}\, dx = -\iint_{\Omega_2} \gamma_2\nabla\tilde{w}^{(1)} \cdot \nabla\overline{\psi}\, dx \quad \text{for all } \psi \in H^1_\diamond(B), \quad (2.18)$$

and $\tilde{w}^{(1)}$ is solution to problem (2.7) for $j = 1$.

Analogously, $S^{(2)} : H^{\frac{1}{2}}_\diamond(\partial\Omega_2) \rightarrow H^{-\frac{1}{2}}_\diamond(\partial\Omega_2)$, $h_2 \mapsto \partial_{\gamma_0}\tilde{v}^{(2)}\big|_{+,2}$ where $\tilde{v}^{(2)} \in H^1_\diamond(B)$ is solution to

$$\iint_B \gamma\nabla\tilde{v}^{(2)} \cdot \nabla\overline{\psi}\, dx = -\iint_{\Omega_1} \gamma_1\nabla\tilde{w}^{(2)} \cdot \nabla\overline{\psi}\, dx \quad \text{for all } \psi \in H^1_\diamond(B), \quad (2.19)$$

and $\tilde{w}^{(2)}$ is solution to (2.7) for $j = 2$. By addition of (2.7) and (2.18) (and of (2.7) and (2.19), respectively) we realize that $\tilde{w}^{(11)} := \tilde{w}^{(1)} + \tilde{v}^{(1)}$ solves (2.16) (and that $\tilde{w}^{(22)} := \tilde{w}^{(2)} + \tilde{v}^{(2)}$ solves (2.17), respectively). We thus conclude that $T^{(1)} + S^{(1)} = T^{(11)}$ and $T^{(2)} + S^{(2)} = T^{(22)}$.

Now it remains to show that $S^{(1)}$ and $S^{(2)}$ are compact operators. We only prove the assertion for $S^{(1)}$ and decompose it as follows: $S^{(1)} = \tilde{S} \circ \hat{S}$, where $\hat{S} : H^{\frac{1}{2}}_\diamond(\partial\Omega_1) \rightarrow H^{\frac{1}{2}}_\diamond(\partial\Omega_2)$, $\tilde{S} : H^{\frac{1}{2}}_\diamond(\partial\Omega_2) \rightarrow H^{-\frac{1}{2}}_\diamond(\partial\Omega_1)$, and

$$h_1 \xrightarrow{\hat{S}} \tilde{w}^{(1)}\big|_{2,-} \xrightarrow{\tilde{S}} \partial_{\gamma_0}\tilde{v}^{(1)}\big|_{+,1},$$

where $\tilde{w}^{(1)}$ is the solution of (2.7) and $\tilde{v}^{(1)}$ solves (2.18). We can apply the very same arguments as in part a) to show that the map $\hat{S}$ is a compact operator. Furthermore, $\tilde{S}$ is obviously bounded, and the proof is completed.

$\square$

Altogether we have derived the representation

$$T = \begin{pmatrix} T^{(1)} & 0 \\ 0 & T^{(2)} \end{pmatrix} + K$$

with a compact operator K. By combination with Lemma 2.1.7 we obtain the following assertion.

Corollary 2.1.9. *Let the operators $T^{(j)}, T_0^{(j)}$ $(j = 1, 2)$ be defined as above. Then the operator $T - T_0$ can be represented as*

$$T - T_0 = \begin{pmatrix} T^{(1)} - T_0^{(1)} & 0 \\ 0 & T^{(2)} - T_0^{(2)} \end{pmatrix} + \tilde{K}$$

with a compact operator $\tilde{K}$

This representation is fundamental for the covering method that is presented in the following section.

2.2 The Covering Method

After the original Factorization method for EIT has been investigated in the previous section we now present our modified version of the method that is capable of detecting mixed inclusions. We first derive the method for real-valued and bounded conductivity contrasts while in Section 2.2.2 we show that we can even extend it to insulating and perfectly conducting inclusions.

2.2.1 Contrasts in the Absolute Conductivity

Let us turn towards the mixed case, i.e. there is an inclusion Ω_1 of type 1 as well as an inclusion Ω_2 of type 2. In this section we derive the Factorization method for slight modifications of $\Lambda - \Lambda_0$ under some additional

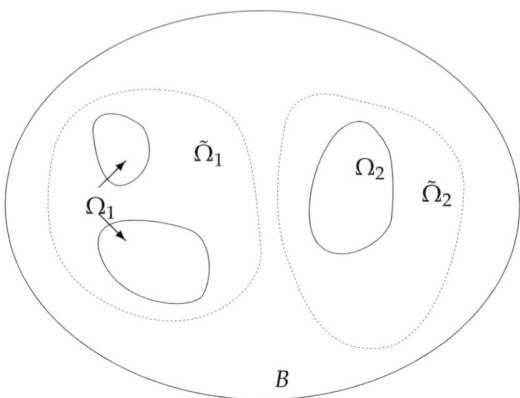

Figure 2.3: Sketch of inclusions and the a priori known domains $\tilde{\Omega}_1, \tilde{\Omega}_2$

a priori assumptions on the inclusions' locations. We show that using these we can reconstruct inclusions even in the mixed case. This setting is defined in the following assumption.

Assumption 2.2.1. We assume that we know C^2-domains $\tilde{\Omega}_1, \tilde{\Omega}_2 \subset B$ for which $\overline{\tilde{\Omega}}_1 \cap \overline{\tilde{\Omega}}_2 = \varnothing$, $\overline{\tilde{\Omega}}_1 \cup \overline{\tilde{\Omega}}_2 \subset B$ and $\overline{\Omega}_j \subset \tilde{\Omega}_j$ $(j = 1, 2)$ hold, and $B \setminus (\overline{\tilde{\Omega}}_1 \cup \overline{\tilde{\Omega}}_2)$ is connected.

This configuration is illustrated in Figure 2.3. The a priori knowledge of $\tilde{\Omega}_1, \tilde{\Omega}_2$ is –at least in some applications– no strong restriction. For example, in medical applications one knows roughly where the different tissues lie and which conductivities they usually have.

According to Lemma 2.1.4 the following coercivity assertions hold for the operators on the diagonal in the representation of $T - T_0$ in Corollary 2.1.9: there exist constants $c_1, c_2 > 0$ such that

$$\left\langle \left(\operatorname{Re} T^{(1)} - T_0^{(1)} \right) h_1, h_1 \right\rangle \geq c_1 \|h_1\|^2_{H^{\frac{1}{2}}(\partial \Omega_1)} \quad \text{for all } h_1 \in H^{\frac{1}{2}}_\diamond (\partial \Omega_1),$$

$$\left\langle - \left(\operatorname{Re} T^{(2)} - T_0^{(2)} \right) h_2, h_2 \right\rangle \geq c_2 \|h_2\|^2_{H^{\frac{1}{2}}(\partial \Omega_2)} \quad \text{for all } h_2 \in H^{\frac{1}{2}}_\diamond (\partial \Omega_2).$$

$$(2.20)$$

Now it is easy to see that the operator $\operatorname{Re} T - T_0$ is neither positively nor negatively coercive, so that we are not able to apply the Factorization method immediately. In order to derive a new factorization for slight modifications of $\Lambda - \Lambda_0$, we need to define some auxiliary operators.

Remark 2.2.2. In order to distinguish between operators that decouple according to the different inclusions and operators that don't decouple we use different notations. Operators with a superscript index j (such as e.g. $T^{(j)}, T_0^{(j)}$) correspond to the case in which Ω_j is considered the only inclusion. On the other hand, we will shortly define several operators with a subscript index (such as e.g. $\tilde{Q}_j$ or $\tilde{G}_j$), and their definition involves both inclusions.

At first let us define $Q^{(1)}, Q^{(2)}$ by

$$Q_j : H_\diamond^{-\frac{1}{2}}(\partial\Omega_j) \to H_\diamond^{-\frac{1}{2}}(\partial\tilde{\Omega}_j), \ g \mapsto \partial_{\gamma_0} w|_{\partial\tilde{\Omega}_j},$$

where $w \in H^1(B \setminus \overline{\Omega}_j)$ with $w|_{\partial B} \in H_\diamond^{\frac{1}{2}}(\partial B)$ solves the boundary value problem

$$\begin{aligned}
\operatorname{div}(\gamma_0 \nabla w) &= 0 \text{ in } B \setminus \overline{\Omega}_j, \\
\partial_{\gamma_0} w &= g \text{ on } \partial\Omega_j, \\
\partial_{\gamma_0} w &= 0 \text{ on } \partial B
\end{aligned} \tag{2.21}$$

in the weak sense ($j = 1, 2$).

Furthermore, we will need the composed operators

$$\tilde{Q}_1 : H_\diamond^{-\frac{1}{2}}(\partial\Omega_1) \times H_\diamond^{-\frac{1}{2}}(\partial\Omega_2) \to H_\diamond^{-\frac{1}{2}}(\partial\tilde{\Omega}_1) \times H_\diamond^{-\frac{1}{2}}(\partial\Omega_2),$$
$$\tilde{Q}_2 : H_\diamond^{-\frac{1}{2}}(\partial\Omega_1) \times H_\diamond^{-\frac{1}{2}}(\partial\Omega_2) \to H_\diamond^{-\frac{1}{2}}(\partial\Omega_1) \times H_\diamond^{-\frac{1}{2}}(\partial\tilde{\Omega}_2).$$

They are defined similarly to the operators $Q^{(1)}, Q^{(2)}$: $\tilde{Q}_1 : (g_1, g_2)^\top \mapsto \left(\partial_{\gamma_0}\tilde{w}|_{\partial\tilde{\Omega}_1}, g_2\right)^\top$, where $\tilde{w} \in H^1(B \setminus \overline{\Omega})$ with $\tilde{w}|_{\partial B} \in H_\diamond^{\frac{1}{2}}(\partial B)$ solves

$$\begin{aligned}
\operatorname{div}(\gamma_0 \nabla \tilde{w}) &= 0 \text{ in } B \setminus \overline{\Omega}, \\
\partial_{\gamma_0}\tilde{w} &= g_j \text{ on } \partial\Omega_j \ (j = 1, 2), \\
\partial_{\gamma_0}\tilde{w} &= 0 \text{ on } \partial B,
\end{aligned} \tag{2.22}$$

in the weak sense, while $\tilde{Q}_2$ is defined by $(g_1, g_2)^\top \mapsto \left(g_1, \partial_{\gamma_0} \tilde{w}|_{\partial \tilde{\Omega}_2}\right)^\top$, where $\tilde{w}$ also solves (2.22). We remark that the boundary value problem (2.22) is the same problem that defines the operator G and thus that the weak formulation corresponding to (2.22) is (2.1).

The following lemma provides some basic properties of the newly defined operators.

Lemma 2.2.3. *Let $Q^{(1)}, Q^{(2)}, \tilde{Q}_1$ and $\tilde{Q}_2$ be defined as above, then:*

(a) *Q_1 and Q_2 are compact operators.*

(b)

$$
\tilde{Q}_1 \left(T - T_0\right) \tilde{Q}_1^* = \begin{pmatrix} Q^{(1)} \left(T^{(1)} - T_0^{(1)}\right) (Q^{(1)})^* & 0 \\ 0 & T^{(2)} - T_0^{(2)} \end{pmatrix}
$$
$$
+ \tilde{K}_1,
$$
$$
\tilde{Q}_2 \left(T - T_0\right) \tilde{Q}_2^* = \begin{pmatrix} T^{(1)} - T_0^{(1)} & 0 \\ 0 & Q^{(2)} \left(T^{(2)} - T_0^{(2)}\right) (Q^{(2)})^* \end{pmatrix}
$$
$$
+ \tilde{K}_2,
$$

$$\tag{2.23}$$

where $\tilde{K}_1, \tilde{K}_2$ are compact operators.

Proof. Part a): Compactness of $Q^{(1)}$ and $Q^{(2)}$ can be shown using the same arguments as in the proof of Lemma 2.1.7.

Part b): We first show that $\tilde{Q}_1$ is a compact perturbation of $\begin{pmatrix} Q^{(1)} & 0 \\ 0 & I \end{pmatrix}$. As in Section 2.1.2 we consider the weak formulations for the boundary value problems (2.21) and (2.22). For (2.21) the weak formulation is to find $w \in H^1(B \setminus \overline{\Omega}_1)$ such that $w|_{\partial B} \in H_\diamond^{\frac{1}{2}}(\partial B)$ and

$$
\iint_{B \setminus \overline{\Omega}_1} \gamma_0 \nabla w \cdot \nabla \overline{\psi} \, dx = \int_{\partial \Omega_1} g_1 \overline{\psi} \, ds \quad \text{for all } \psi \in H^1(B \setminus \overline{\Omega}_1), \tag{2.24}
$$

while for (2.22) we have to find $\tilde{w} \in H^1(B \setminus \overline{\Omega})$ such that $\tilde{w}|_{\partial B} \in H^{\frac{1}{2}}_\diamond(\partial B)$ and

$$\iint_{B \setminus \overline{\Omega}} \gamma_0 \nabla \tilde{w} \cdot \nabla \overline{\psi}\, dx = \int_{\partial \Omega_1} g_1 \overline{\psi}\, ds + \int_{\partial \Omega_2} g_2 \overline{\psi}\, ds \quad \text{for all } \psi \in H^1(B \setminus \overline{\Omega}),$$

Setting $v := w - \tilde{w}$ in $B \setminus \overline{\Omega}$ we obtain by subtraction and the first Green identity that

$$\iint_{B \setminus \overline{\Omega}} \gamma_0 \nabla v \cdot \nabla \overline{\psi}\, dx = -\iint_{\Omega_2} \gamma_0 \nabla w \cdot \nabla \overline{\psi}\, dx - \int_{\partial \Omega_2} g_2 \overline{\psi}\, ds$$

$$= -\int_{\partial \Omega_2} (\partial_{\gamma_0} w + g_2)\, \overline{\psi}\, ds$$

for all $\psi \in H^1(B \setminus \Omega_1)$. Now we observe that the difference operator $\hat{Q} := \begin{pmatrix} Q^{(1)} & 0 \\ 0 & I \end{pmatrix} - \tilde{Q}_1$ maps $(g_1, g_2)^\top$ to $\left(\partial_{\gamma_0} v|_{\partial \tilde{\Omega}_1}, 0\right)^\top$. Furthermore, the map

$$\hat{Q}_1 : H^{-\frac{1}{2}}_\diamond(\partial \Omega_1) \times H^{-\frac{1}{2}}_\diamond(\partial \Omega_2) \to H^{-\frac{1}{2}}_\diamond(\partial \Omega_2),$$

$$(g_1, g_2)^\top \mapsto \partial_{\gamma_0} w|_{+,2} + g_2,$$

where w solves (2.24) is a bounded linear operator. Using the same arguments as in the proof of Lemma 2.1.8 we can show that the map

$$\hat{Q}_2 : H^{-\frac{1}{2}}_\diamond(\partial \Omega_2) \to H^{-\frac{1}{2}}_\diamond(\partial \tilde{\Omega}_1) \times H^{-\frac{1}{2}}_\diamond(\partial \Omega_2), \quad g \mapsto (\partial_{\gamma_0} v, 0)^\top,$$

where $v \in H^1(B \setminus \overline{\Omega})$ with $v|_{\partial B} \in H^{\frac{1}{2}}_\diamond(\partial B)$ solves

$$\iint_{B \setminus \overline{\Omega}} \gamma_0 \nabla v \cdot \nabla \overline{\psi}\, dx = -\int_{\partial \Omega_2} g \overline{\psi}\, ds \quad \text{for all } \psi \in H^1(B \setminus \overline{\Omega})$$

is a compact operator. By construction it is clear that $\hat{Q} = \hat{Q}_2 \circ \hat{Q}_1$ holds which implies that there is a representation $\tilde{Q}_1 = \begin{pmatrix} Q^{(1)} & 0 \\ 0 & I \end{pmatrix} + K$ with a compact operator K. Since $Q^{(1)}$ is also compact, plugging this representation into $\tilde{Q}_1(T - T_0)\tilde{Q}_1^*$ yields the assertion. The proof for the second equality is analogous.

$\square$

In addition, we will need the operators $\tilde{G}^{(j)} : H_\diamond^{-\frac{1}{2}}(\partial\tilde{\Omega}_j) \to L_\diamond^2(\partial B)$ $(j = 1, 2)$ that are defined just as G in Section 2.1 but with Ω replaced by $\tilde{\Omega}_j$, i.e. the operator $\tilde{G}^{(j)}$ maps $g_j \mapsto v|_{\partial B}$, where $v \in H^1(B \setminus \overline{\tilde{\Omega}}_j)$ with $v|_{\partial B} \in H_\diamond^{\frac{1}{2}}(\partial B)$ is the weak solution to the boundary value problem

$$\text{div}(\gamma_0 \nabla v) = 0 \text{ in } B \setminus \overline{\tilde{\Omega}}_j,$$
$$\partial_{\gamma_0} v = g_j \text{ on } \partial\tilde{\Omega}_j, \qquad (2.25)$$
$$\partial_{\gamma_0} v = 0 \text{ on } \partial B.$$

Furthermore, we define $\tilde{G}_1 : H_\diamond^{-\frac{1}{2}}(\partial\tilde{\Omega}_1) \times H_\diamond^{-\frac{1}{2}}(\partial\tilde{\Omega}_2) \to L_\diamond^2(\partial B)$ by $(g_1, g_2)^\top \mapsto \tilde{v}|_{\partial B}$, and $\tilde{v} \in H^1(B \setminus (\overline{\tilde{\Omega}}_1 \cup \overline{\Omega}_2))$ with $\tilde{v}|_{\partial B} \in H_\diamond^{\frac{1}{2}}(\partial B)$ solves

$$\text{div}(\gamma_0 \nabla \tilde{v}) = 0 \text{ in } B \setminus (\overline{\tilde{\Omega}}_1 \cup \overline{\Omega}_2),$$
$$\partial_{\gamma_0} \tilde{v} = g_1 \text{ on } \partial\tilde{\Omega}_1,$$
$$\partial_{\gamma_0} \tilde{v} = g_2 \text{ on } \partial\Omega_2,$$
$$\partial_{\gamma_0} \tilde{v} = 0 \text{ on } \partial B$$

in the weak sense. For the operator $\tilde{G}_2 : H_\diamond^{-\frac{1}{2}}(\partial\Omega_1) \times H_\diamond^{-\frac{1}{2}}(\partial\tilde{\Omega}_2) \to L_\diamond^2(\partial B)$ the corresponding definition holds.

It is quite obvious that $\tilde{G}^{(1)}, \tilde{G}^{(2)}, \tilde{G}_1, \tilde{G}_2$ have the same properties as G concerning compactness, injectivity, and denseness of the range (compare page 16). The following Lemma shows some connections between the newly defined operators.

Lemma 2.2.4. *The operators defined above have the following properties, where* K_1, K_2 *are compact operators:*

(a)

$$G = \tilde{G}_1 \tilde{Q}_1, \quad G = \tilde{G}_2 \tilde{Q}_2,$$

(b)

$$\tilde{G}^{(1)}(\tilde{G}^{(1)})^* = \tilde{G}_1 \begin{pmatrix} I & K_1^* \\ K_1 & K_1 K_1^* \end{pmatrix} \tilde{G}_1^*$$

$$\tilde{G}^{(2)}(\tilde{G}^{(2)})^* = \tilde{G}_2 \begin{pmatrix} K_2 K_2^* & K_2 \\ K_2^* & I \end{pmatrix} \tilde{G}_2^*.$$

(c) $\tilde{Q}_1$ and $\tilde{Q}_2$ are one-to-one and have dense range. The same holds for their adjoints $\tilde{Q}_1^*$ and $\tilde{Q}_2^*$

Proof. Part a) follows immediately from the definition of the appearing operators.

Part b): Define $K_1 : H_\diamond^{-\frac{1}{2}}(\partial\tilde{\Omega}_1) \to H_\diamond^{-\frac{1}{2}}(\partial\Omega_2)$ by $g_1 \mapsto \partial_{\gamma_0} v|_{\partial\Omega_2}$ and $v \in H^1(B \setminus \overline{\tilde{\Omega}}_1)$ with $v|_{\partial B} \in H_\diamond^{\frac{1}{2}}(\partial B)$ solves (2.25) for $j = 1$ in the weak sense. Analogously to the proof of Lemma 2.1.8 it can be shown that K_1 is compact.

We observe that the identity $\tilde{G}^{(1)} = \tilde{G}_1 \begin{pmatrix} I & 0 \\ K_1 & 0 \end{pmatrix}$ holds which implies $(\tilde{G}^{(1)})^* = \begin{pmatrix} I & K_1^* \\ 0 & 0 \end{pmatrix} \tilde{G}_1^*$ and thus $\tilde{G}^{(1)}(\tilde{G}^{(1)})^* = \tilde{G}_1 \begin{pmatrix} I & K_1^* \\ K_1 & K_1 K_1^* \end{pmatrix} \tilde{G}_1^*$. The arguments for the second identity are analogous.

Part c): Since G is injective we deduce from part a) that $\tilde{Q}_1$ is injective and thus that $\tilde{Q}_1^*$ has dense range in $H_\diamond^{-\frac{1}{2}}(\partial\Omega_1) \times H_\diamond^{-\frac{1}{2}}(\partial\Omega_2)$.

Now we derive a representation for the adjoint $\tilde{Q}_1^*$ and show that it is injective. We therefore first show an alternative representation for $\tilde{Q}_1$.

Consider the partial operator $\tilde{Q}^{(11)} : H_\diamond^{-\frac{1}{2}}(\partial\Omega_1) \to H_\diamond^{-\frac{1}{2}}(\partial\tilde{\Omega}_1)$ that maps $g_1 \mapsto \partial_{\gamma_0} w_1|_{\partial\tilde{\Omega}_1}$, and $w_1 \in H^1(B \setminus \overline{\Omega})$ with $w_1|_{\partial B} \in H_\diamond^{\frac{1}{2}}(\partial B)$ solves $\operatorname{div}(\gamma_0 \nabla w_1) = 0$ in $B \setminus \overline{\Omega}$, $\partial_{\gamma_0} w_1 = g_1$ on $\partial\Omega_1$, $\partial_{\gamma_0} w_1 = 0$ on $\partial\Omega_2 \cup \partial B$ in the weak sense.

In addition, consider the operator $\tilde{Q}^{(12)} : H_\diamond^{-\frac{1}{2}}(\partial\Omega_2) \to H_\diamond^{-\frac{1}{2}}(\partial\tilde{\Omega}_1)$ that maps $g_2 \mapsto \partial_{\gamma_0} w_2|_{\partial\tilde{\Omega}_1}$, and $w_2 \in H^1(B \setminus \overline{\Omega})$ with $w_2|_{\partial B} \in H_\diamond^{\frac{1}{2}}(\partial B)$ solves $\operatorname{div}(\gamma_0 \nabla w_2) = 0$ in $B \setminus \overline{\Omega}$, $\partial_{\gamma_0} w_2 = g_2$ on $\partial\Omega_2$, $\partial_{\gamma_0} w_2 = 0$ on $\partial\Omega_1 \cup \partial B$ in the weak sense.

Now we observe that $w := w_1 + w_2$ solves (2.22) and thus that the identity $\tilde{Q}_1 = \begin{pmatrix} \tilde{Q}^{(11)} & \tilde{Q}^{(12)} \\ 0 & I \end{pmatrix}$ holds, which also implies $\tilde{Q}_1^* = \begin{pmatrix} (\tilde{Q}^{(11)})^* & 0 \\ (\tilde{Q}^{(12)})^* & I \end{pmatrix}$. The next step is to show representations for the adjoints $(\tilde{Q}^{(11)})^*$, $(\tilde{Q}^{(12)})^*$.

In fact, $(\tilde{Q}^{(11)})^* : H_\diamond^{\frac{1}{2}}(\partial\tilde{\Omega}_1) \to H_\diamond^{\frac{1}{2}}(\partial\Omega_1)$, $h_1 \mapsto v|_{\partial\Omega_1}$, and $v \in H^1(B \setminus (\Omega \cup \partial\tilde{\Omega}_1))$ with $v|_{\partial B} \in H_\diamond^{\frac{1}{2}}(\partial B)$ solves the transmission boundary value problem $\operatorname{div}(\gamma_0 \nabla v) = 0$ in $B \setminus (\Omega \cup \partial\tilde{\Omega}_1)$, $\partial_{\gamma_0} v = 0$ on $\partial B \cup \partial\Omega_1 \cup \partial\Omega_2$, $v|_- - v|_+ = h_1$ in $\partial\tilde{\Omega}_1$ and $\partial_{\gamma_0} v|_- = \partial_{\gamma_0} v|_+$ on $\partial\tilde{\Omega}_1$ in the weak sense.

Now we can use the second Green identity to show that $\tilde{Q}^{(11)}$ and $(\tilde{Q}^{(11)})^*$ are indeed adjoint to each other. Therefore let $g_1 \in H_\diamond^{-\frac{1}{2}}(\partial\Omega_1)$ and $h_1 \in H_\diamond^{\frac{1}{2}}(\partial\tilde{\Omega}_1)$ be arbitrary, and let w_1, v be the corresponding weak solutions of the above boundary value problems, then:

$$\left\langle \tilde{Q}^{(11)} g_1, h_1 \right\rangle = \int_{\partial\tilde{\Omega}_1} \partial_{\gamma_0} w_1 (\bar{v}|_- - \bar{v}|_+) - w_1 \underbrace{(\partial_{\gamma_0}\bar{v}|_- - \partial_{\gamma_0}\bar{v}|_+)}_{=0}\, ds$$

$$= \int_{\partial\Omega_1} \bar{v}\partial_{\gamma_0} w_1 - w_1 \underbrace{\partial_{\gamma_0}\bar{v}}_{=0}\, ds + \int_{\partial\Omega_2} \bar{v}\underbrace{\partial_{\gamma_0} w_1}_{=0} - w_1 \underbrace{\partial_{\gamma_0}\bar{v}}_{=0}\, ds$$

$$- \int_{\partial B} \bar{v}\underbrace{\partial_{\gamma_0} w_1}_{=0} - w_1 \underbrace{\partial_{\gamma_0}\bar{v}}_{=0}\, ds = \left\langle g_1, (\tilde{Q}^{(11)})^* h_1 \right\rangle.$$

In the same way it can be shown that the adjoint of $\tilde{Q}^{(12)}$ is given by $(\tilde{Q}^{(12)})^* : H_\diamond^{\frac{1}{2}}(\partial\tilde{\Omega}_1) \to H_\diamond^{\frac{1}{2}}(\partial\Omega_2)$, $h_1 \mapsto v|_{\partial\Omega_2}$, where v is defined by the above transmission problem.

The only thing that remains to show is that $\tilde{Q}_1^*$ is injective. Therefore let $(h_1, h_2) \in H_\diamond^{\frac{1}{2}}(\partial\tilde{\Omega}_1) \times H_\diamond^{\frac{1}{2}}(\partial\Omega_2)$ be such that $(\tilde{Q}^{(11)})^* h_1 = 0$ and $(\tilde{Q}^{(12)})^* h_1 + h_2 = 0$. The first identity yields that v has zero Cauchy values at $\partial\Omega_1$ and thus that $v \equiv 0$ in $\tilde{\Omega}_1 \setminus \overline{\Omega}_1$. In addition, the continuity of $\partial_{\gamma_0} v$ across $\partial\tilde{\Omega}_1$ yields that v solves the homogeneous Neumann problem in $B \setminus (\overline{\Omega}_1 \cup \overline{\Omega}_2)$ and thus that $v \equiv 0$ in the whole of B which also implies $h_1 = 0$. Now the second identity yields $h_2 = 0$ immediately, and injectivity of $\tilde{Q}_1^*$ is shown, which also implies that $\tilde{Q}_1$ has dense range in $H_\diamond^{-\frac{1}{2}}(\partial\tilde{\Omega}_1) \times H_\diamond^{-\frac{1}{2}}(\partial\Omega_2)$. For $\tilde{Q}_2$ the arguments are completely analogous. $\qquad\square$

Figures 2.4 and 2.5 illustrate the connections between the factorization (2.5) and the newly introduced operators $\tilde{G}_1$, $\tilde{G}_2$, $\tilde{Q}_1$ and $\tilde{Q}_2$ that have

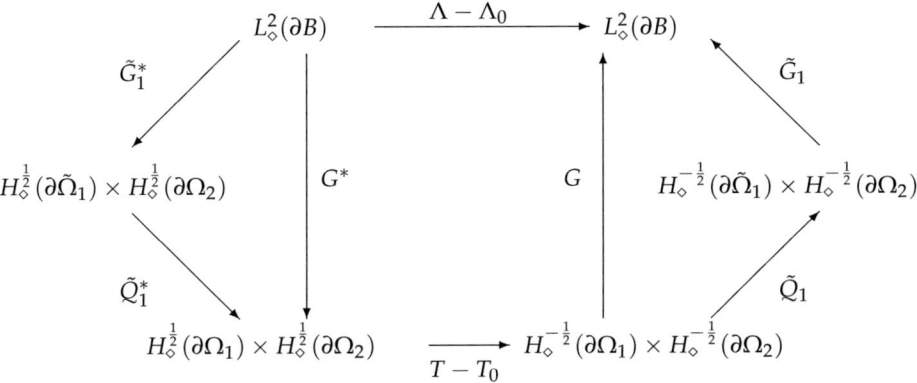

Figure 2.4: Diagram of operators and spaces corresponding to the new factorization of $\Lambda - \Lambda_0$ (compare Lemma 2.2.4 part a))

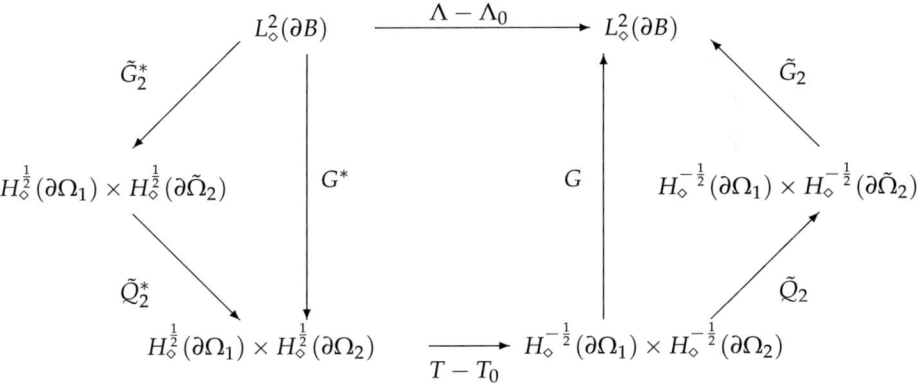

Figure 2.5: Diagram of operators and spaces corresponding to the new factorization of $\Lambda - \Lambda_0$ (compare Lemma 2.2.4 part b))

been shown in Lemma 2.2.4. Now we have collected all the tools we need to derive our new factorization.

Therefore consider the following perturbed Neumann-to-Dirichlet difference maps:

$$\tilde{\Lambda}_1 := \Lambda - \Lambda_0 + \rho_1 \tilde{G}^{(1)}(\tilde{G}^{(1)})^*, \quad \tilde{\Lambda}_2 := \Lambda - \Lambda_0 + \rho_2 \tilde{G}^{(2)}(\tilde{G}^{(2)})^*,$$

with parameters $\rho_1, \rho_2 \in \mathbb{C}$. First, we remark that both maps only contain measured data and information known a priori: Λ is obtained from measurement data, Λ_0 is obtained either from measurement data or from information about the background conductivity γ_0 and for $\tilde{G}_j, \tilde{G}_j^*$ we only need information about γ_0 and the covering domains $\tilde{\Omega}_j$ ($j = 1, 2$). Using Lemma 2.2.4 and the factorization (2.5) we can derive factorizations of $\tilde{\Lambda}_1$ and $\tilde{\Lambda}_2$:

$$\tilde{\Lambda}_1 = G(T - T_0)G^* + \rho_1 \tilde{G}^{(1)}(\tilde{G}^{(1)})^*$$
$$= \tilde{G}_1 \left\{ \tilde{Q}_1(T - T_0)\tilde{Q}_1^* + \rho_1 \begin{pmatrix} I & K_1^* \\ K_1 & K_1 K_1^* \end{pmatrix} \right\} \tilde{G}_1^*,$$

$$\tilde{\Lambda}_2 = G(T - T_0)G^* + \rho_2 \tilde{G}^{(2)}(\tilde{G}^{(2)})^*$$
$$= \tilde{G}_2 \left\{ \tilde{Q}_2(T - T_0)\tilde{Q}_2^* + \rho_2 \begin{pmatrix} K_2 K_2^* & K_2 \\ K_2^* & I \end{pmatrix} \right\} \tilde{G}_2^*.$$

Our aim is now to show that for these factorizations the Factorization method works. On order to do this, several properties of the appearing operators have to be verified. These properties can be established by choosing the parameters ρ_1, ρ_2 in a proper way as we will see in the following considerations.

We therefore proceed by investigating the new middle operators S_1, S_2 defined by

$$S_1 = \tilde{Q}_1 (T - T_0) \tilde{Q}_1^* + \rho_1 \begin{pmatrix} I & K_1^* \\ K_1 & K_1 K_1^* \end{pmatrix}, \qquad (2.26)$$

$$S_2 = \tilde{Q}_2 (T - T_0) \tilde{Q}_2^* + \rho_2 \begin{pmatrix} K_2 K_2^* & K_2 \\ K_2^* & I \end{pmatrix}. \qquad (2.27)$$

Lemma 2.2.5. *Let $\rho_1, \rho_2 \in \mathbb{C}$ be such that $\operatorname{Re}\rho_1 < 0$, $\operatorname{Re}\rho_2 > 0$ and $\operatorname{Im}\rho_j > 0$ ($j = 1, 2$). Then the middle operators S_1, S_2 have the following properties:*

(a) $\operatorname{Im}\langle S_1 h, h\rangle \geq 0$ *for all* $h = (h_1, h_2)^\top \in H_\diamond^{\frac{1}{2}}(\partial\tilde{\Omega}_1) \times H_\diamond^{\frac{1}{2}}(\partial\Omega_2)$ *and*
   $\operatorname{Im}\langle S_2 h, h\rangle \geq 0$ *for all* $h = (h_1, h_2)^\top \in H_\diamond^{\frac{1}{2}}(\partial\Omega_1) \times H_\diamond^{\frac{1}{2}}(\partial\tilde{\Omega}_2)$.

(b) If $0 \neq h = (h_1, h_2)^\top$ is such that $(\operatorname{Re} S_j)h = 0$, then $\operatorname{Im}\langle S_j h, h\rangle > 0$ ($j = 1, 2$).

Proof. Part a): We only consider the operator S_1 from (2.26) while for S_2 the arguments are the same:

$$S_1 = \underbrace{\tilde{Q}_1 (T - T_0) \tilde{Q}_1^*}_{=:H} + \rho_1 \underbrace{\begin{pmatrix} I & K_1^* \\ K_1 & K_1 K_1^* \end{pmatrix}}_{=:\tilde{H}}.$$

From (2.6) we know that $\langle \operatorname{Im} Hh, h\rangle \geq 0$ for all $h \in H_\diamond^{\frac{1}{2}}(\partial\tilde{\Omega}_1) \times H_\diamond^{\frac{1}{2}}(\partial\Omega_2)$, and $\tilde{H}$ is non-negative since it is an operator of the type A^*A. Hence

$$\operatorname{Im}\langle S_1 h, h\rangle = \langle \operatorname{Im} Hh, h\rangle + \operatorname{Im}\{\rho_1\}\langle \tilde{H}h, h\rangle \geq 0,$$

and the first assertion is shown.

Part b): Let $h = (h_1, h_2)^\top \in H_\diamond^{\frac{1}{2}}(\partial\tilde{\Omega}_1) \times H_\diamond^{\frac{1}{2}}(\partial\Omega_2)$ satisfy $(\operatorname{Re} S_1)h = 0$ and $h \neq 0$. If $\operatorname{Re} Hh = 0$ then it follows that $\tilde{H}h = 0$. From Lemmas 2.1.5 and 2.2.4 we obtain that $\langle \operatorname{Im} Hh, h\rangle > 0$ and thus $\langle \operatorname{Im} S_1 h, h\rangle > 0$. If $\operatorname{Re} Hh \neq 0$, then we conclude that $\tilde{H}h \neq 0$ and thus $\langle \tilde{H}h, h\rangle > 0$. This implies together with part a) that $\operatorname{Im}\langle S_1 h, h\rangle > 0$ which completes the proof.

$\square$

Now we derive another representation of $\tilde{\Lambda}_1$ that is used to prove that the middle operator S_1 is a compact perturbation of a negatively coercive operator. Using Corollary 2.1.9 as well as Lemmas 2.2.3 and 2.2.4 we obtain

$$
\begin{aligned}
\tilde{\Lambda}_1 &= G \left\{ \begin{pmatrix} T^{(1)} - T_0^{(1)} & 0 \\ 0 & T^{(2)} - T_0^{(2)} \end{pmatrix} + \tilde{K} \right\} G^* + \rho_1 \tilde{G}^{(1)} (\tilde{G}^{(1)})^* \\
&= \tilde{G}_1 \left\{ \begin{pmatrix} \rho_1 I & 0 \\ 0 & T^{(2)} - T_0^{(2)} \end{pmatrix} \right. \\
&\quad \left. + \begin{pmatrix} Q^{(1)}(T^{(1)} - T_0^{(1)})(Q^{(1)})^* & K_1^* \\ K_1 & 0 \end{pmatrix} + \tilde{K} \right\} \tilde{G}_1^* \\
&= \tilde{G}_1 \left\{ \begin{pmatrix} \rho_1 I & 0 \\ 0 & T^{(2)} - T_0^{(2)} \end{pmatrix} + \tilde{K}_1 \right\} \tilde{G}_1^*.
\end{aligned}
\tag{2.28}
$$

$\tilde{K}_1$ is a compact operator, and if the parameter ρ_1 is chosen as in Lemma 2.2.5 then from (2.20) it follows that $\operatorname{Re} S_1$ is indeed a compact perturbation of a negatively coercive operator. Analogously we can derive a corresponding representation for $\tilde{\Lambda}_2$:

$$
\begin{aligned}
\tilde{\Lambda}_2 &= G \left\{ \begin{pmatrix} T^{(1)} - T_0^{(1)} & 0 \\ 0 & T^{(2)} - T_0^{(2)} \end{pmatrix} + \tilde{K} \right\} G^* + \tilde{G}^{(2)} (\tilde{G}^{(2)})^* \\
&= \tilde{G}_2 \left\{ \begin{pmatrix} T^{(1)} - T_0^{(1)} & 0 \\ 0 & \rho_2 I \end{pmatrix} \right. \\
&\quad \left. + \begin{pmatrix} 0 & K_2 \\ K_2^* & Q^{(2)}(T^{(2)} - T_0^{(2)})(Q^{(2)})^* \end{pmatrix} + \tilde{K} \right\} \tilde{G}_2^* \\
&= \tilde{G}_2 \left\{ \begin{pmatrix} T^{(1)} - T_0^{(1)} & 0 \\ 0 & \rho_2 I \end{pmatrix} + \tilde{K}_2 \right\} \tilde{G}_2^*,
\end{aligned}
\tag{2.29}
$$

with a compact operator $\tilde{K}_2$, and choosing ρ_2 as in Lemma 2.2.5 and (2.20) yield that $\operatorname{Re} S_2$ is a compact perturbation of a coercive operator.

Now we summarize our results from (2.28) and (2.29) as well as from Lemma 2.2.5 and the knowledge about the properties of the operators $\tilde{G}_1$ and $\tilde{G}_2$. We have derived the two factorizations

$$
\tilde{\Lambda}_1 = \tilde{G}_1 S_1 \tilde{G}_2^*, \quad \tilde{\Lambda}_2 = \tilde{G}_2 S_2 \tilde{G}_2^*,
$$

and the appearing operators have the following properties:

(a) $\tilde{\Lambda}_1 : L^2_\diamond(\partial B) \to L^2_\diamond(\partial B)$ and $\tilde{\Lambda}_2 : L^2_\diamond(\partial B) \to L^2_\diamond(\partial B)$ are bounded operators.

(b) The operators $\tilde{G}_1 : H_\diamond^{-\frac{1}{2}}(\partial\tilde{\Omega}_1) \times H_\diamond^{-\frac{1}{2}}(\partial\Omega_2) \to L^2_\diamond(\partial B)$ and $S_1 : H_\diamond^{\frac{1}{2}}(\partial\tilde{\Omega}_1) \times H_\diamond^{\frac{1}{2}}(\partial\Omega_2) \to H_\diamond^{-\frac{1}{2}}(\partial\tilde{\Omega}_1) \times H_\diamond^{-\frac{1}{2}}(\partial\Omega_2)$ are bounded.

(c) The operators $\tilde{G}_2 : H_\diamond^{-\frac{1}{2}}(\partial\Omega_1) \times H_\diamond^{-\frac{1}{2}}(\partial\tilde{\Omega}_2) \to L^2_\diamond(\partial B)$ and $S_2 : H_\diamond^{\frac{1}{2}}(\partial\Omega_1) \times H_\diamond^{\frac{1}{2}}(\partial\tilde{\Omega}_2) \to H_\diamond^{-\frac{1}{2}}(\partial\Omega_1) \times H_\diamond^{-\frac{1}{2}}(\partial\tilde{\Omega}_2)$ are bounded.

(d) $\tilde{G}_1$ and $\tilde{G}_2$ are compact operators and they have dense range in $L^2_\diamond(\partial B)$.

(e) The operator $\operatorname{Re} S_1$ can be written as $\operatorname{Re} S_1 = \mathcal{C}_1 + \mathcal{K}_1$ where $\mathcal{K}_1$ is compact, $\mathcal{C}_1$ is self-adjoint and $-\mathcal{C}_1$ is coercive.

(f) The operator $\operatorname{Re} S_2$ can be written as $\operatorname{Re} S_2 = \mathcal{C}_2 + \mathcal{K}_2$ where $\mathcal{K}_2$ is compact and $\mathcal{C}_2$ is self-adjoint and coercive.

(g) The operator $\operatorname{Im} S_1$ is non-negative on $H_\diamond^{\frac{1}{2}}(\partial\tilde{\Omega}_1) \times H_\diamond^{\frac{1}{2}}(\partial\Omega_2)$, and the operator $\operatorname{Im} S_2$ is non-negative on $H_\diamond^{\frac{1}{2}}(\partial\Omega_1) \times H_\diamond^{\frac{1}{2}}(\partial\tilde{\Omega}_2)$.

(h) $\operatorname{Im} S_j$ is positive on the finite-dimensional nullspace of $\operatorname{Re} S_j$ ($j = 1, 2$).

These results yield that we can apply the $F_\#$-Factorization Method in the form of the following theorem that is cited from [63].

Theorem 2.2.6. *Let $X \subset U \subset X^*$ be a Gelfand triple with a Hilbert space U and a reflexive Banach space X such that the imbedding is dense. Let H be a second Hilbert space and let $\mathcal{F} : H \to H, \mathcal{G} : X^* \to H$ and $\mathcal{S} : X \to X^*$ be linear and bounded operators with $\mathcal{F} = \mathcal{G}\mathcal{S}\mathcal{G}^*$ and*

(a) *$\mathcal{G}$ is one-to-one and compact with dense range in H,*

(b) *$\operatorname{Re} \mathcal{S} = \mathcal{C} + \mathcal{K}$ where $\mathcal{C}$ (or $-\mathcal{C}$) is coercive and $\mathcal{K}$ is compact,*

(c) *$\operatorname{Im} \mathcal{S}$ is non-negative on X,*

(d) *$\mathcal{S}$ is injective or $\operatorname{Im} \mathcal{S}$ is positive on the finite-dimensional nullspace of $\operatorname{Re} \mathcal{S}$.*

Then $\mathcal{F}_{\#} := |\operatorname{Re} \mathcal{F}| + \operatorname{Im} \mathcal{F}$ is positive definite and the ranges of $\mathcal{G} : X^* \to H$ and $\mathcal{F}_{\#}^{\frac{1}{2}} : H \to H$ coincide.

For the corresponding proof we refer to [63]. As a result, we conclude for our factorizations:

Corollary 2.2.7. *The operators $\Lambda_{\#,1}, \Lambda_{\#,2} : L^2_\diamond(\partial B) \to L^2_\diamond(\partial B)$ defined by*

$$\Lambda_{\#,1} = \left|\operatorname{Re} \Lambda - \Lambda_0 + \operatorname{Re}\{\rho_1\}\tilde{G}^{(1)}(\tilde{G}^{(1)})^*\right| + \operatorname{Im} \Lambda + \operatorname{Im}\{\rho_1\}\tilde{G}^{(1)}(\tilde{G}^{(1)})^*,$$

$$\Lambda_{\#,2} = \left|\operatorname{Re} \Lambda - \Lambda_0 + \operatorname{Re}\{\rho_2\}\tilde{G}^{(2)}(\tilde{G}^{(2)})^*\right| + \operatorname{Im} \Lambda + \operatorname{Im}\{\rho_2\}\tilde{G}^{(2)}(\tilde{G}^{(2)})^*,$$

are positive definite, and the ranges of $\tilde{G}_1$ and $\Lambda_{\#,1}^{\frac{1}{2}}$ as well as the ranges of $\tilde{G}_2$ and $\Lambda_{\#,2}^{\frac{1}{2}}$ coincide.

By means of the test functions φ_y defined in (2.12) we can again derive a binary criterion to decide whether a point y lies inside an inclusion or not. Therefore we first show a connection between the inclusions and the ranges of $\tilde{G}_1$ and $\tilde{G}_2$.

Lemma 2.2.8. *The inclusions Ω_1, Ω_2 can be characterized as follows:*

a) Let $y \in \tilde{\Omega}_1$, then $y \in \Omega_1 \Leftrightarrow \varphi_y \in \mathcal{R}(\tilde{G}_2)$.

b) Let $y \in \tilde{\Omega}_2$, then $y \in \Omega_2 \Leftrightarrow \varphi_y \in \mathcal{R}(\tilde{G}_1)$.

Proof. We only prove assertion a). Under the assumption $y \in \tilde{\Omega}_1$ the equivalence $y \in \Omega_1 \Leftrightarrow y \in \Omega_1 \cup \tilde{\Omega}_2$ holds, since $\tilde{\Omega}_1$ and $\tilde{\Omega}_2$ are disjoint. Additionally, we know that $y \in \Omega_1 \cup \tilde{\Omega}_2 \Leftrightarrow \varphi_y \in \mathcal{R}(\tilde{G}_2)$ since it is the very same result as for the original Factorization method (compare (2.13)). $\square$

Note that the assumptions $y \in \tilde{\Omega}_j$ $(j = 1,2)$ are no restriction in the present setting since in Assumption 2.2.1 we require that we a priori know that the inclusions are located inside the corresponding covering domains. By combination of Lemma 2.2.8 and Corollary 2.2.7 we obtain a binary criterion that connects the data $\Lambda_{\#,1}, \Lambda_{\#,2}$ with the desired inclusions Ω_1, Ω_2.

Corollary 2.2.9. *Let Ω_1 be of type 1, let Ω_2 be of type 2 and let Assumption 2.2.1 be satisfied.*

a) Let $y \in \tilde{\Omega}_1$, then

$$y \in \Omega_1 \Leftrightarrow \sum_{k=1}^{\infty} \frac{|\langle \varphi_y, \psi_k \rangle|^2}{\lambda_k} < \infty, \qquad (2.30)$$

where $\{\lambda_k, \psi_k : k \in \mathbb{N}\}$ is an eigensystem of $\Lambda_{\#,2}$.

b) Let $y \in \tilde{\Omega}_2$, then

$$y \in \Omega_2 \Leftrightarrow \sum_{k=1}^{\infty} \frac{|\langle \varphi_y, \psi_k \rangle|^2}{\lambda_k} < \infty, \qquad (2.31)$$

where $\{\lambda_k, \psi_k : k \in \mathbb{N}\}$ is an eigensystem of $\Lambda_{\#,1}$.

These two binary criterions can now be used to reconstruct inclusions. However, in contrast to the original Factorization method we now have to implement two steps in order to identify both inclusions. In the first step we use criterion (2.30) and reconstruct Ω_1, while in the second step we use criterion (2.31) to reconstruct Ω_2.

The new method can also be interpreted as follows: for the reconstruction of Ω_1 we use the parameter ρ_2 to synthesize a type 1 inclusion $\tilde{\Omega}_2$ that covers the perturbing type 2 inclusion Ω_2. On the other hand, we cover Ω_1 by an artificial type 2 inclusion that is produced by ρ_1 in order to identify Ω_2.

2.2.2 Insulating and Perfectly Conducting Inclusions

In our statements above perfectly conducting as well as insulating inclusions have been excluded since we assumed the absolute conductivity $\operatorname{Re} \gamma$ to be strictly positive and essentially bounded. In this section we show that these two special cases also fit in our setting. In particular, we show that for insulating inclusions the middle operator of the factorization is coercive, while for perfectly conducting inclusions the middle operator is negatively coercive.

At first we consider the factorization for the case of an insulating inclusion and and no other inclusions. If Ω is an insulating inclusion, then there is no current flux across its boundary, i.e. the direct problem is to find a solution $u \in H^1(B \setminus \overline{\Omega})$ with $u|_{\partial B} \in H_{\diamond}^{\frac{1}{2}}(\partial B)$ to

$$\iint_{B \setminus \overline{\Omega}} \gamma_0 \nabla u \cdot \nabla \overline{\psi} \, dx = \int_{\partial B} f \overline{\psi} \, ds \quad \text{for all } \psi \in H^1(B \setminus \overline{\Omega}),$$

where $f \in L_{\diamond}^2(\partial B)$ is the injected current pattern. This weak formulation corresponds to the following Neumann boundary value problem

$$\text{div}\, (\gamma_0 \nabla u) = 0 \text{ in } B \setminus \overline{\Omega},$$
$$\partial_{\gamma_0} u = 0 \text{ on } \partial \Omega,$$
$$\partial_{\gamma_0} u = f \text{ on } \partial B.$$

The homogeneous Neumann boundary condition at $\partial \Omega$ yields that

$$(\Lambda - \Lambda_0) f = G \left(- \partial_{\gamma_0} u_0|_+ \right),$$

where $u_0 \in H_{\diamond}^1(B)$ solves the background direct problem. Now we can define the operator $L : L_{\diamond}^2(\partial B) \rightarrow H_{\diamond}^{-\frac{1}{2}}(\partial \Omega)$, $f \mapsto \partial_{\gamma_0} u_0|_{\partial \Omega}$ for which the identity $\Lambda - \Lambda_0 = GL$ holds. Using the second Green identity it can be shown that the adjoint $L^* : H_{\diamond}^{\frac{1}{2}}(\partial \Omega) \rightarrow L_{\diamond}^2(\partial B)$ is such that it maps $h \mapsto w_0|_{\partial B}$, where w_0 solves (2.3) in the weak sense for the background case $\gamma \equiv \gamma_0$:

$$\langle Lf, h \rangle = \int_{\partial \Omega} \partial_{\gamma_0} u_0 (\overline{w}_0|_+ - \overline{w}_0|_-) - u_0 (\underbrace{\partial_{\gamma_0} \overline{w}_0|_+ - \partial_{\gamma_0} \overline{w}_0|_-}_{=0}) \, ds$$

$$= \int_{\partial B} \overline{w}_0 \partial_{\gamma_0} u_0 - u_0 \underbrace{\partial_{\gamma_0} \overline{w}_0}_{=0} \, ds = \langle f, L^* h \rangle.$$

Now we observe that $L^* = -GT_0$ and thus that $L = -T_0 G^*$. On the whole, we obtain the following factorization for Ω being an insulator:

$$\Lambda - \Lambda_0 = G(-T_0)G^*.$$

The desired coercivity property for the middle operator $-T_0$ can now be shown as follows. Let $h \in H_{\diamond}^{\frac{1}{2}}(\partial \Omega)$ and let $w_0 \in H^1(B \setminus \overline{\Omega})$ be the corresponding weak solution of (2.3) for $\gamma \equiv \gamma_0$, then

$$\langle -T_0 h, h \rangle_{\partial\Omega} = -\int_{\partial\Omega} \partial_{\gamma_0} w_0|_+ \left(\overline{w}_0|_+ - \overline{w}_0|_- \right) ds$$

$$= \iint_{B\setminus\overline{\Omega}} \gamma_0 |\nabla w_0|^2 \, dx - \int_{\partial B} \overline{w}_0 \, \partial_{\gamma_0} w_0 \, ds + \iint_{\Omega} \gamma_0 |\nabla w_0|^2 \, dx$$

$$= \iint_B \gamma_0 |\nabla w_0|^2 \, dx \geq c_0 \iint_B |\nabla w_0|^2 \, dx,$$

where $c_0 > 0$ is the lower bound for γ_0 from Assumption 2.1.2. Now we can argue as in the proof of Theorem 2.3 in [53] to obtain that there is $c > 0$ such that

$$\langle -T_0 h, h \rangle_{\partial\Omega} \geq c \, \|h\|^2_{H^{\frac{1}{2}}(\partial\Omega)} \quad \text{for all } h \in H^{\frac{1}{2}}_\diamond(\partial\Omega).$$

Since the middle operator is positively coercive we know that an insulating inclusion can be treated in our covering method exactly the same as an inclusion of type 1.

Now we consider the second case. Let Ω be a perfect conductor which is roughly described by $\gamma|_\Omega = \infty$. For the sake of simplicity we restrict ourselves to the case in which the perfectly conducting inclusion Ω consists of only one component. For the case of several components we refer to [2] or [8]. A perfect conductor Ω is modeled by the claim that the electric potential inside Ω has to be constant. The direct problem for a given current pattern $f \in L^2_\diamond(\partial B)$ is then to find $u \in H^1(B \setminus \overline{\Omega})$ with $u|_{\partial B} \in H^{\frac{1}{2}}_\diamond(\partial B)$ satisfying $u = const$ on $\partial\Omega$ and

$$\iint_{B\setminus\overline{\Omega}} \gamma_0 \nabla u \cdot \nabla \overline{\psi} \, dx = \int_{\partial B} f \overline{\psi} \, ds$$

for all $\psi \in H^1(B \setminus \overline{\Omega})$ such that $\psi|_{\partial\Omega} = const$. The corresponding classical formulation is

$$\text{div}\,(\gamma_0 \nabla u) = 0 \text{ in } B \setminus \overline{\Omega},$$
$$\partial_{\gamma_0} u = f \text{ on } \partial B, \tag{2.32}$$
$$u = const \text{ on } \partial\Omega.$$

Remark 2.2.10. The constant in the boundary condition $u|_{\partial \Omega} = const$ is determined uniquely by the claim that $u|_{\partial B} \in H_\diamond^{\frac{1}{2}}(\partial B)$. We could alternatively formulate a well-posed direct problem for $u \in H^1(B \setminus \overline{\Omega})$ using the homogeneous Dirichlet boundary condition $u|_{\partial \Omega} = 0$. However, since we wish to find a factorization for the difference operator $\Lambda - \Lambda_0$, it is desirable to have $u|_{\partial B} \in H_\diamond^{\frac{1}{2}}(\partial B)$.

As in the insulator case above we now derive a factorization of $\Lambda - \Lambda_0$ and show negative coercivity of the middle operator. From the direct problem we observe that the identity $\Lambda = G\hat{L}$ holds with the operator $\hat{L} : L_\diamond^2(\partial B) \to H_\diamond^{-\frac{1}{2}}(\partial \Omega)$ that maps $f \mapsto \partial_{\gamma_0} u|_{\partial \Omega}$, where $u \in H^1(B \setminus \overline{\Omega})$ with $u|_{\partial B} \in H_\diamond^{\frac{1}{2}}(\partial B)$ solves (2.32) in the weak sense. The adjoint of $\hat{L}$ is given by $\hat{L}^* : H_\diamond^{\frac{1}{2}}(\partial \Omega) \to L_\diamond^2(\partial B)$ mad maps $h \mapsto w|_{\partial B}$, where $w \in H^1(B \setminus \overline{\Omega})$ with $w|_{\partial B} \in H_\diamond^{\frac{1}{2}}(\partial B)$ is the weak solution to

$$\mathrm{div}\,(\gamma_0 \nabla w) = 0 \text{ in } B \setminus \overline{\Omega},$$
$$\partial_{\gamma_0} w = 0 \text{ on } \partial B,$$
$$w = h + const \text{ on } \partial \Omega.$$

The proof that $\hat{L}, \hat{L}^*$ are adjoint to each other is again based on the second Green identity:

$$\langle \hat{L}f, h \rangle = \int_{\partial \Omega} \overline{w} \partial_{\gamma_0} u \, ds - \underbrace{\int_{\partial \Omega} u \partial_{\gamma_0} \overline{w} \, ds}_{=0}$$

$$= \int_{\partial B} \overline{w} \partial_{\gamma_0} u - u \underbrace{\partial_{\gamma_0} \overline{w}}_{=0} \, ds = \langle f, \hat{L}^* h \rangle.$$

Now we define the operator $\hat{T} : H_\diamond^{\frac{1}{2}}(\partial \Omega) \to H_\diamond^{-\frac{1}{2}}(\partial \Omega)$ by setting $h \mapsto \partial_{\gamma_0} w|_{\partial \Omega}$, and the factorization

$$\Lambda - \Lambda_0 = G(\hat{T} - T_0)G^*$$

is shown.

In order to show negative coercivity of the middle operator we use the first Green identity:

$$\langle \hat{T}h, h \rangle_{\partial\Omega} = \int_{\partial\Omega} \overline{w} \, \partial_{\gamma_0} w - const \, \partial_{\gamma_0} w \, ds$$

$$= \int_{\partial B} \overline{w} \, \partial_{\gamma_0} w \, ds - \iint_{B\backslash\overline{\Omega}} \gamma_0 \, |\nabla w|^2 \, dx = - \iint_{B\backslash\overline{\Omega}} \gamma_0 \, |\nabla w|^2 \, dx$$

and thus

$$\langle (T_0 - \hat{T})h, h \rangle_{\partial\Omega} = \iint_{B\backslash\overline{\Omega}} \gamma_0 \, |\nabla w|^2 \, dx - \iint_{B} \gamma_0 \, |\nabla w_0|^2 \, dx.$$

We extend w to a function $w \in H^1(B \setminus \partial\Omega)$ by setting $w = const$ inside Ω and obtain

$$\langle (T_0 - \hat{T})h, h \rangle_{\partial\Omega} = \iint_{B} \gamma_0 \, |\nabla w|^2 - \gamma_0 \, |\nabla w_0|^2 \, dx$$

$$= -2 \iint_{B} \gamma_0 \nabla w_0 \cdot (\nabla \overline{w}_0 - \nabla \overline{w}) \, dx$$

$$+ \iint_{B} \gamma_0 \, |\nabla w_0|^2 - 2\gamma_0 \nabla w_0 \cdot \nabla \overline{w} + \gamma_0 \, |\nabla w|^2 \, dx.$$

Here the first integral vanishes which can be seen by setting $\psi = w_0 - w$ in (2.2) for the background case $\gamma = \gamma_0$. Hence we can estimate

$$\langle (T_0 - T)h, h \rangle_{\partial\Omega} = \iint_{B\backslash\overline{\Omega}} \gamma_0 \, |\nabla w_0 - \nabla w|^2 \, dx + \iint_{\Omega} \gamma_0 \, |\nabla w_0|^2 \, dx$$

$$\geq \iint_{\Omega} \gamma_0 \, |\nabla w_0|^2 \, dx.$$

Again we can argue as in the proof of Theorem 2.3 in [53] to show coercivity of $T_0 - \hat{T}$. Hence we can treat Ω in the same manner as a type 2 inclusion in our covering method.

Altogether we have shown in this section that for the covering method we can subsume all type 1 inclusions together with the insulators and cover it by an a priori known covering domain $\tilde{\Omega}_1$. Analogously, we can subsume all the type 2 inclusions together with the perfect conductors and cover them by $\tilde{\Omega}_2$. We can detect the type 1 inclusions and insulators using the binary criterion from part b) of Corollary 2.2.9, and we can detect all type 2 inclusions and perfect conductors using part a) of Corollary 2.2.9, respectively.

2.3 Numerical Experiments

In this section we present some results of numerical experiments with our new method. We compare the results to those obtained using the original Factorization method and investigate its performance depending on the parameters ρ_1, ρ_2 as well as the covering domains $\tilde{\Omega}_1, \tilde{\Omega}_2$.

2.3.1 The Original Factorization Method

In all our examples the domain B is the unit disc in $\mathbb{R}^2$ and the background conductivity is $\gamma_0 \equiv 1$. In the upper half of B there is a type 1 inclusion Ω_1, while in the lower half of B there is a type 2 inclusion Ω_2. The conductivity in B is as follows:

$$\gamma(x) = \begin{cases} 0.5, & x \in \Omega_1, \\ 2, & x \in \Omega_2, \\ 1, & x \in B \setminus \overline{\Omega}. \end{cases}$$

We consider two examples with different locations and shapes of the inclusions Ω_1 and Ω_2. In the first example Ω_1 is kite-shaped and lies in the upper half of B, while Ω_2 is a small ellipse and is located in the bottom right of B. In the second example both Ω_1 and Ω_2 are peanut-shaped and are located in the center of the upper and lower halves of B. In Figure 2.6 these two examples are illustrated.

Since B is the unit disc, the dipole test function φ_y can be represented explicitly by

$$\varphi_y(x) = \frac{1}{2\pi} \frac{(y - x) \cdot \hat{a}}{|y - x|^2},$$

where $\hat{a}$ is a fixed unit vector that represents the dipole axis (see e.g. [53]). As an orthonormal basis for the current patterns as well as the boundary

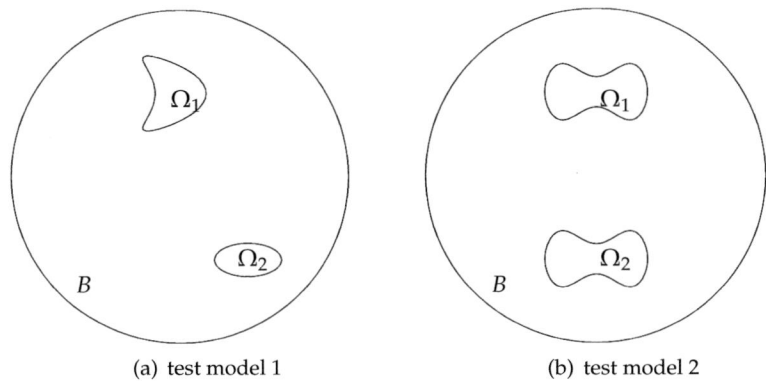

(a) test model 1   (b) test model 2

Figure 2.6: The two test models and with the inclusions Ω_1, Ω_2

potentials we use the trigonometric functions

$$\left\{ \frac{1}{\sqrt{\pi}} \cos(k\theta), \frac{1}{\sqrt{\pi}} \sin(k\theta) : k = 1, 2, \dots \right\}, \tag{2.33}$$

where θ denotes the argument of points on the boundary ∂B in polar coordinates.

In order to obtain discrete approximations of Λ and Λ_0 we solve the direct problem with and without the inclusions for the basis current patterns up to an end index $N \in \mathbb{N}$ using a standard finite element method.

For the reconstruction of the inclusions, we evaluate the functions

$$W_j(y) = \left(\sum_{k=1}^{N} \frac{\left\langle \varphi_y, \psi_k^{(j)} \right\rangle^2}{\lambda_k^{(j)}} \right)^{-1} \quad (j = 1, 2)$$

on a mesh of points y in B. As stated in Corollary 2.2.9, we obtain the reconstruction of Ω_1 by evaluation of W_2, where $\{ \lambda_k^{(2)}, \psi_k^{(2)} : k \in \mathbb{N} \}$ is an eigensystem of $\Lambda_{\#,2}$. On the other hand, we can reconstruct Ω_2 by evaluating W_1 on a mesh, where $\{ \lambda_k^{(1)}, \psi_k^{(1)} : k \in \mathbb{N} \}$ is an eigensystem

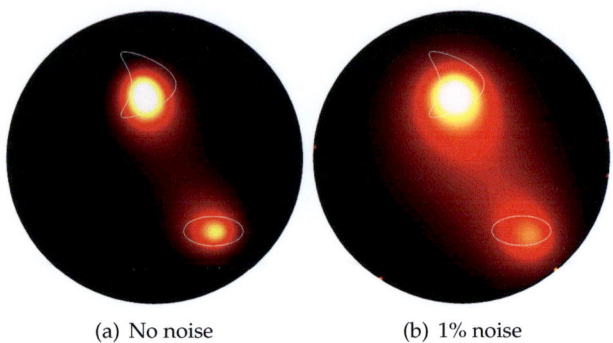

(a) No noise   (b) 1% noise

Figure 2.7: Reconstruction for test model 1 using the standard Factorization
method

of $\Lambda_{\#,1}$. In [61] a justification for the truncation of the Picard series at the
index N can be found.

For our first reconstructions we set $\rho_1 = \rho_2 = 0$, which implies that
$\Lambda_{\#,1} = \Lambda_{\#,2}$ and thus $W_1 = W_2$. This corresponds to the standard Factorization method for EIT which is not proven to work for the present exam-
ples. Figure 2.7 shows the reconstruction for Example 1, for the noiseless
case (left picture) and for 1% of white noise added to our discrete approximation of $\Lambda - \Lambda_0$. In Figure 2.8 the corresponding reconstructions for
Example 2 are shown.

From these examples we observe that even the standard Factorization
method seems to be capable of reconstructing inclusions in the mixed
case. This observation was also made in Figure 7 in [46] and, for scattering problems, in [37].

2.3.2 Different Parameters and Noise Levels

The experience from scattering theory (see e.g. [37]) also suggests that
the original Factorization method produces better reconstructions than
the covering method although it is not proven to work in the case of
mixed obstacles.

Furthermore, the larger the absolute values of the parameters ρ_1, ρ_2

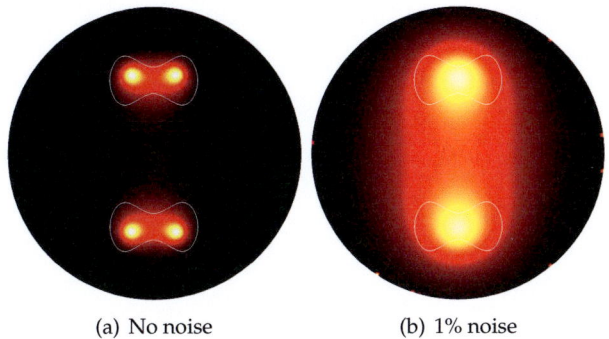

(a) No noise   (b) 1% noise

Figure 2.8: Reconstruction for test model 2 using the standard Factorization method

get, the more they might influence the reconstruction as we can see in our theory: the larger $|\rho_j|$ is ($j = 1, 2$) the larger the contrast of the artificial covering inclusion to the background gets, and the Factorization method emphasizes this synthetic inclusion rather than the one that we wish to identify. This can also be observed in the representations (2.26), (2.27) of the middle operator. The information about the desired inclusion lies inside the first part

$$\tilde{Q}_1 \left(T - T_0\right) \tilde{Q}_1^* \ \text{ or } \ \tilde{Q}_2 \left(T - T_0\right) \tilde{Q}_2^*,$$

respectively, while the second part

$$\rho_1 \begin{pmatrix} I & K_1^* \\ K_1 & K_1 K_1^* \end{pmatrix} \ \text{ or } \ \rho_2 \begin{pmatrix} K_2 K_2^* & K_2 \\ K_2^* & I \end{pmatrix},$$

respectively, contains information about the covering domains $\tilde{\Omega}_1, \tilde{\Omega}_2$ and carries more weight as $|\rho_j|$ ($j = 1, 2$) gets larger.

In the following numerical examples we verify these expectations for our new method where we carried out both reconstructions using different values of the parameters ρ_1, ρ_2.

For the first examples with the covering method the covering domains $\tilde{\Omega}_1, \tilde{\Omega}_2$ are circles as shown in Figure 2.9. The domains in which we evaluate the indicator functions W_j ($j = 1, 2$) are ellipses as indicated by the

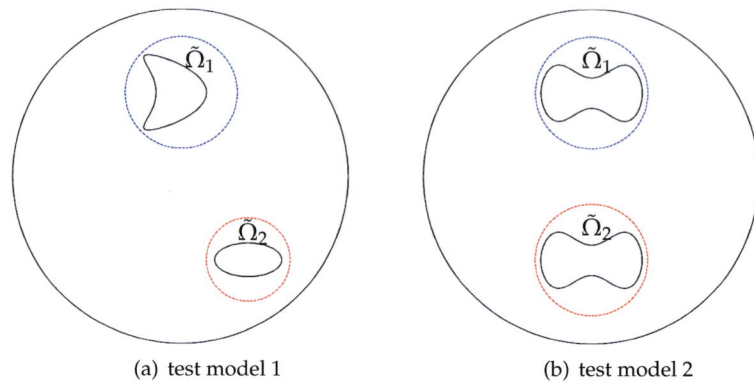

(a) test model 1   (b) test model 2

Figure 2.9: Covering domains for the test models

white dotted lines in the reconstructions. We start by the reconstructions in the noiseless case and different values of ρ_1 and ρ_2.

Figures 2.10 and 2.11 show the corresponding reconstructions. We observe that the inclusions are localized properly by our method but the reconstructions are not as focused as with the original Factorization method (compare Figures 2.7 and 2.8). The absolute value of the parameters ρ_1, ρ_2 also has the expected influence on the reconstruction quality: the results for $\rho_1 = 10^{-3}(-1+i)$ and $\rho_2 = 10^{-3}(1+i)$ (parts a) and c) of the Figures 2.10 and 2.11) are considerably better than those for $\rho_1 = 10^{-2}(-1+i)$ and $\rho_2 = 10^{-2}(1+i)$ (parts b) and d) of Figures 2.10 and 2.11).

We proceed with the same examples for noisy data. In the Figures 2.12 and 2.13 the results of the very same experiment are shown but with 1% white noise added to $\Lambda - \Lambda_0$. The reconstructions are slightly more blurred compared to the noiseless case. However, the observations concerning the absolute value of ρ_1, ρ_2 as well as the quality compared to the standard Factorization method are the same.

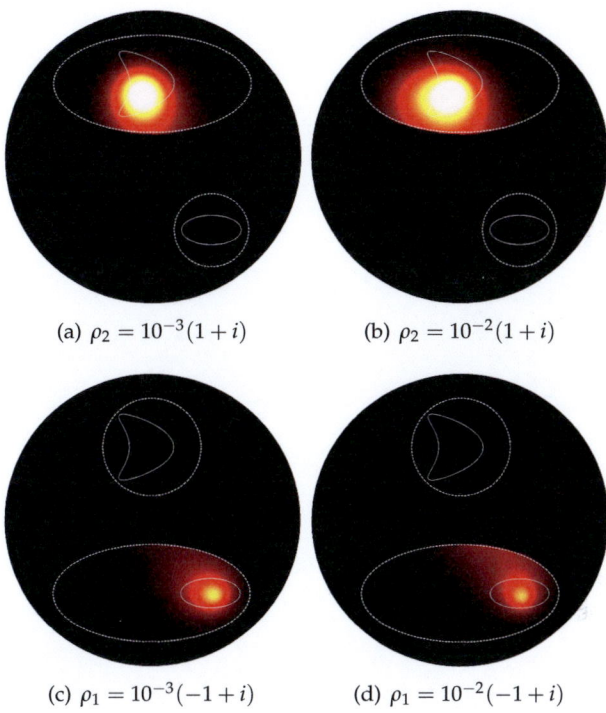

(a) $\rho_2 = 10^{-3}(1+i)$ (b) $\rho_2 = 10^{-2}(1+i)$

(c) $\rho_1 = 10^{-3}(-1+i)$ (d) $\rho_1 = 10^{-2}(-1+i)$

Figure 2.10: Reconstruction for test model 1 using the covering method

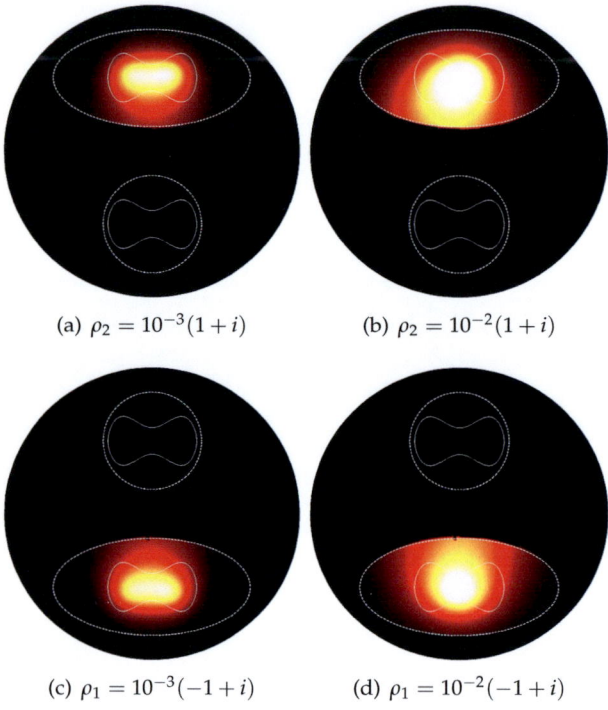

(a) $\rho_2 = 10^{-3}(1+i)$ (b) $\rho_2 = 10^{-2}(1+i)$

(c) $\rho_1 = 10^{-3}(-1+i)$ (d) $\rho_1 = 10^{-2}(-1+i)$

Figure 2.11: Reconstruction for test model 2 using the covering method

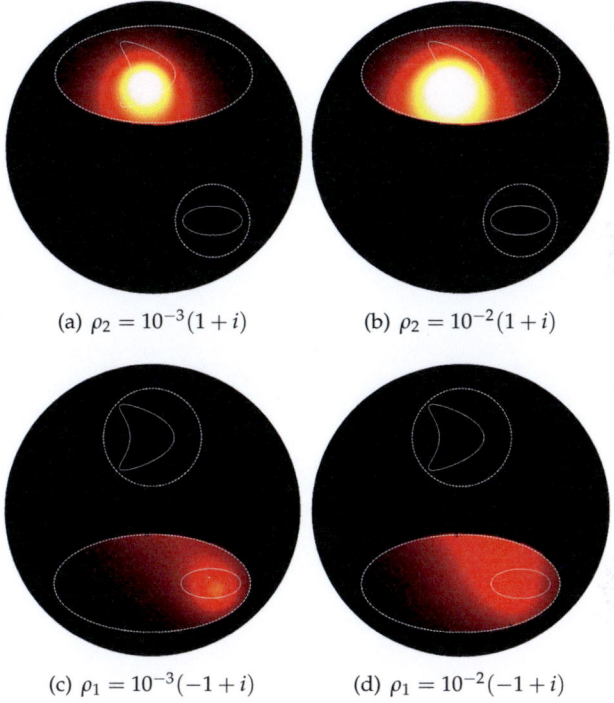

(a) $\rho_2 = 10^{-3}(1+i)$

(b) $\rho_2 = 10^{-2}(1+i)$

(c) $\rho_1 = 10^{-3}(-1+i)$

(d) $\rho_1 = 10^{-2}(-1+i)$

Figure 2.12: Reconstruction for test model 1 using the covering method; 1% white noise added to $\Lambda - \Lambda_0$

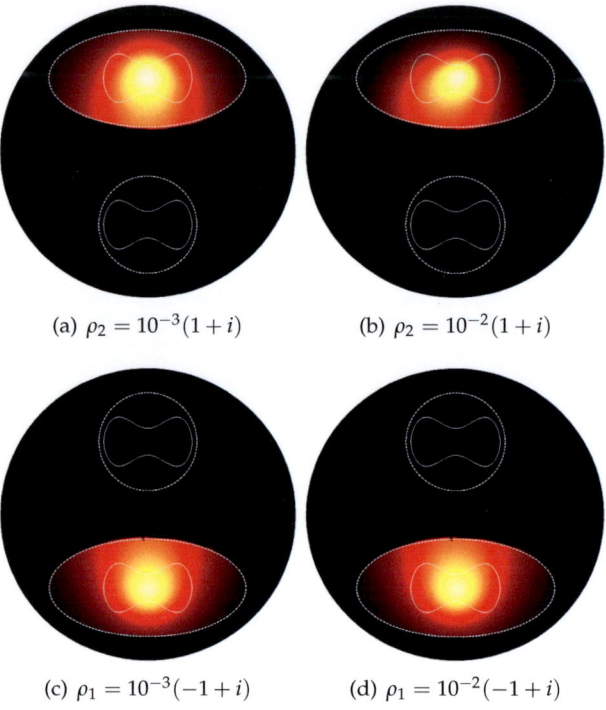

(a) $\rho_2 = 10^{-3}(1+i)$ (b) $\rho_2 = 10^{-2}(1+i)$

(c) $\rho_1 = 10^{-3}(-1+i)$ (d) $\rho_1 = 10^{-2}(-1+i)$

Figure 2.13: Reconstruction for test model 2 using the covering method; 1% white noise added to $\Lambda - \Lambda_0$

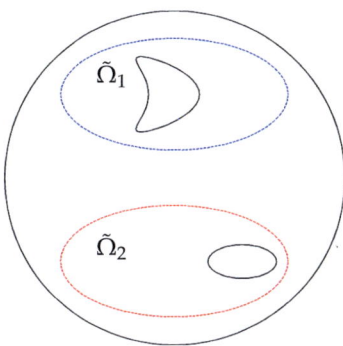

Figure 2.14: Large covering domains for test model 1

2.3.3 Different Covering Domains

Now we investigate the influence of the size of the covering domains $\tilde{\Omega}_1, \tilde{\Omega}_2$ on our reconstructions. In the examples from above they were chosen relatively small, hence we now choose larger coverings as they are illustrated in Figure 2.14. For the following examples we restrict to the test model 1 since the results for the test model 2 are very similar.

We observe that they are noticeably worse than the corresponding reconstructions for the same values of ρ_1, ρ_2 and the previous smaller covering domains. This observation can be explained as follows: since the synthetic inclusion caused by the covering method is now larger than before it is more disturbing for the localization of the desired inclusion. It is thus recommended to choose the coverings as small as possible.

Our last experiment deals with the case in which the covering domains do not cover the inclusions properly, i.e. the assumed a priori information is incorrect. For the choice of the covering domains $\tilde{\Omega}_1, \tilde{\Omega}_1$ compare Figure 2.16.

The reconstructions for this case are shown in Figure 2.17. They are slightly worse than the reconstructions in Figure 2.10 but still have an edge on the ones in Figure 2.15 where the coverings were larger. This observation is in some sense plausible since the unions $\tilde{\Omega}_j \cup \Omega_j$ ($j = 1, 2$) are only slightly larger than the coverings in Figure 2.10 but still much smaller than the elliptic coverings before. However, we have to emphasize that this case is not dealt with by our theory.

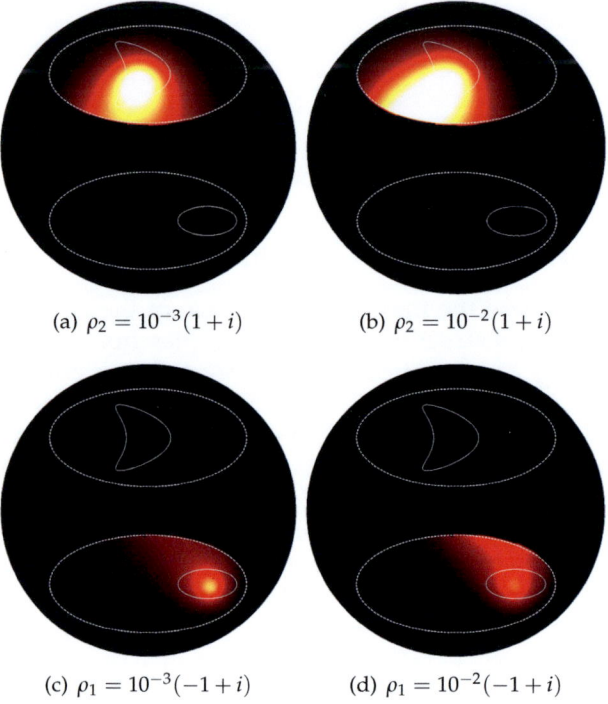

(a) $\rho_2 = 10^{-3}(1+i)$ (b) $\rho_2 = 10^{-2}(1+i)$

(c) $\rho_1 = 10^{-3}(-1+i)$ (d) $\rho_1 = 10^{-2}(-1+i)$

Figure 2.15: Reconstruction for test model 1 using the covering method and large ellipses as covering domains

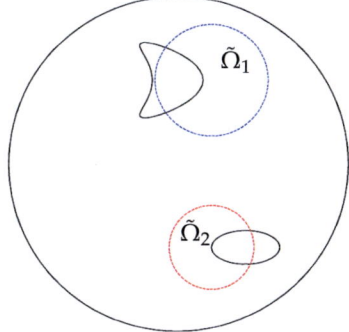

Figure 2.16: Improper covering domains for the test model 1

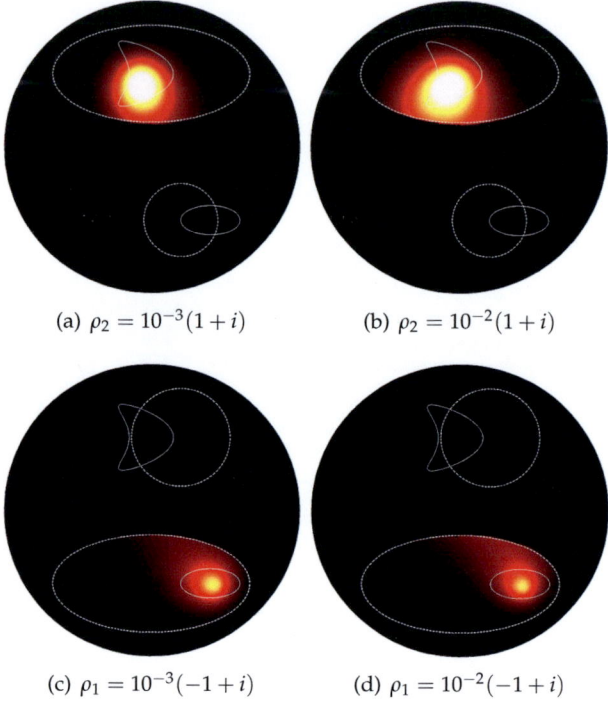

(a) $\rho_2 = 10^{-3}(1+i)$ (b) $\rho_2 = 10^{-2}(1+i)$

(c) $\rho_1 = 10^{-3}(-1+i)$ (d) $\rho_1 = 10^{-2}(-1+i)$

Figure 2.17: Reconstruction for test model 1 using the covering method and improper covering domains

3 Determination of the Conductivity Inside Anomalies

The Factorization method for EIT is a very successful method for the localization of anomalies. However, one possible drawback of this method is that the conductivity inside the inclusions remains unknown. The aim of this chapter is to derive a method for approximating the conductivity inside inclusions after they have been identified by a qualitative method, such as e.g. the Factorization method. This plan is comparable to the considerations in [12], [14] and [13], where the authors present methods of approximating the surface impedance or the index of refraction of a scatterer whose shape has been determined before by a qualitative method.

The main idea of our method is to make use of a factorization of $\Lambda - \Lambda_0$ that appears in a new version of the Factorization method which has been derived by Kirsch in [55]. Our method of determining the conductivity is –in contrast to the assumptions in Chapter 2– restricted to the case of piecewise constant conductivities. Furthermore, we emphasize that this problem is still an ill-posed problem.

In Section 3.1 we present the new factorization and show substantial properties of the appearing operators. In Section 3.2 we focus on the middle operator T of the factorization and especially its spectrum that is closely related to the conductivity inside the inclusions. Afterwards, in Section 3.3, we show how this spectrum and the conductivity can be computed numerically and in Section 3.4 we present some numerical results with this new method.

3.1 A new Version of the Factorization Method

Our assumptions on the domain B and the inclusion Ω from the previous chapter are still valid, i.e. B is a simply connected C^2-domain, Ω is a C^2-domain such that $\overline{\Omega} \subset B$ and $B \setminus \overline{\Omega}$ is connected. For the conductivity distribution we restrict ourselves to piecewise constant and real-valued conductivities, i.e. $\gamma(x) = \gamma_0 + q\chi_\Omega(x)$ ($x \in B$), and $\gamma_0, q \in \mathbb{R}$ are such that $\gamma(x) > 0$ in B and $q \neq 0$. In Section 1.2.1 we noticed that the mean value of the absolute conductivity in B is equal to 1. Hence we can assume that $\gamma_0 = 1$ without loss of generality. Later we will extend our results to complex-valued q as well as several conductivity contrasts $q_1, \ldots, q_n$ on N mutually disjoint inclusions $\Omega_1, \ldots, \Omega_n$.

We proceed by defining the operators that will appear in our new factorization and start with the operator $\mathcal{A} : L^2_\diamond(\partial B) \to L^2(\Omega, \mathbb{R}^d)$, defined by $f \mapsto \nabla u_0|_\Omega$, where $u_0 \in H^1_\diamond(B)$ solves

$$\iint_B \nabla u_0 \cdot \nabla \psi \, dx = \int_{\partial B} f\psi \, ds \quad \text{for all } \psi \in H^1_\diamond(B). \tag{3.1}$$

This weak formulation corresponds to the classical formulation

$$\Delta u_0 = 0 \text{ in } B, \quad \partial_\nu u_0 = f \text{ on } \partial B. \tag{3.2}$$

The term ∂_ν is a short notation for the normal derivative $\frac{\partial}{\partial \nu}$. Hence $\mathcal{A}$ is defined via the background direct problem. $\mathcal{A}^* : L^2(\Omega, \mathbb{R}^d) \to L^2_\diamond(\partial B)$ is given by $h \mapsto w|_{\partial B}$ where $w \in H^1_\diamond(B)$ solves

$$\iint_B \nabla w \cdot \nabla \psi \, dx = \iint_\Omega h \cdot \nabla \psi \, dx \quad \text{for all } \psi \in H^1_\diamond(B). \tag{3.3}$$

$\mathcal{A}$ and $\mathcal{A}^*$ are adjoint to each other since the first Green identity yields together with (3.3) that

$$\langle \mathcal{A}f, h \rangle_{L^2(\Omega, \mathbb{R}^d)} = \iint_\Omega \nabla u_0 \cdot h \, dx = \iint_B \nabla u_0 \cdot \nabla w \, dx = \langle f, \mathcal{A}^*h \rangle_{L^2(\partial B)}$$

holds for arbitrary $f \in L^2_\diamond(\partial B), h \in L^2(\Omega, \mathbb{R}^d)$. Using the unique continuation principle for harmonic functions it can be shown that $\mathcal{A}$ is injective which also implies that $\mathcal{A}^*$ has dense range in $L^2_\diamond(\partial B)$.

In the following lemma we show representations for the closure of the range of $\mathcal{A}$ and the nullspace of its adjoint $\mathcal{A}^*$. We therefore need to declare what we mean by $\Delta v = 0$ in Ω for a function $v \in H^1(\Omega)$, namely:

$$\iint_\Omega \nabla v \cdot \nabla \psi \, dx = 0 \quad \text{for all } \psi \in H^1_0(\Omega). \tag{3.4}$$

Lemma 3.1.1. *The operators $\mathcal{A}, \mathcal{A}^*$ defined above have the following properties:*

(a) *The nullspace $\mathcal{N}(\mathcal{A}^*)$ consists of all $h \in L^2(\Omega, \mathbb{R}^d)$ such that there is a solution $w \in H^1(\Omega)$ to*

$$\iint_{\Omega} \nabla w \cdot \nabla \psi \, dx = \iint_{\Omega} h \cdot \nabla \psi \, dx \quad \text{for all } \psi \in H^1(\Omega) \qquad (3.5)$$

satisfying $w = 0$ on $\partial \Omega$.

(b)
$$\overline{\mathcal{R}(\mathcal{A})} = \left\{ \nabla u : u \in H^1(\Omega), \, \Delta u = 0 \text{ in } \Omega \right\}. \qquad (3.6)$$

Proof. Part a): Let $h \in \mathcal{N}(\mathcal{A}^*)$. Then there is $w \in H^1(B)$ that solves (3.3) and $w = 0$ on ∂B. This implies that w has zero Cauchy values on ∂B and thus $w \equiv 0$ in $B \setminus \overline{\Omega}$ by the unique continuation principle. For such a w (3.3) is equivalent to (3.5). Furthermore, since $w \in H^1(B)$, it is continuous across $\partial \Omega$ and therefore $w|_{\partial \Omega} = 0$.

Let $h \in L^2(\Omega, \mathbb{R}^d)$ be such that there is a solution $w \in H^1(\Omega)$ to (3.5) and $w|_{\partial \Omega} = 0$. Then we can extend w to $w \in H^1_\diamond(B)$ by setting $w \equiv 0$ in $B \setminus \overline{\Omega}$. Now w is a solution to (3.3), and $w|_{\partial B} = 0$. Hence $h \in \mathcal{N}(\mathcal{A}^*)$.

Part b): Define $M := \left\{ \nabla u : u \in H^1(\Omega), \, \Delta u = 0 \text{ in } \Omega \right\}$. From the definition of $\mathcal{A}$ we obtain $\mathcal{R}(\mathcal{A}) \subset M$. From (3.4) it follows immediately that M is a closed subspace of $H^1(\Omega)$ and thus that $\overline{\mathcal{R}(\mathcal{A})} \subset M$.

Let $\nabla v \in M$ and let $h \in \mathcal{N}(\mathcal{A}^*)$. Then we know from part a) that there exists $w \in H^1(\Omega)$ that solves (3.5) and $w|_{\partial \Omega} = 0$. Hence we obtain using (3.5) and (3.4) that

$$\iint_{\Omega} \nabla v \cdot h \, dx = \iint_{\Omega} \nabla v \cdot \nabla w \, dx = 0.$$

This implies that $M \subset (\mathcal{N}(\mathcal{A}^*))^{\perp} = \overline{\mathcal{R}(\mathcal{A})}$, which completes the proof. $\square$

Remark 3.1.2. (a) The classical formulation for the variational problem (3.5) is to find $w \in H^1(\Omega)$ that solves $\Delta w = \text{div } h$ in Ω, $w = 0$ on $\partial \Omega$ and $\partial_\nu w = h \cdot \nu$ on $\partial \Omega$.

(b) Lemma 3.1.1 implies in particular that $\mathcal{A}^*$ is not one-to-one. For instance, let us fix some arbitrary point $\hat{x}$ inside Ω and and choose $\epsilon > 0$ such that $dist(\hat{x}, \partial\Omega) > \epsilon$. Then we define

$$w(x) := \begin{cases} \exp\left(1 - \dfrac{\epsilon^2}{\epsilon^2 - |x - \hat{x}|^2}\right), & |x - \hat{x}| < \epsilon, \\ 0, & \text{otherwise.} \end{cases}$$

We observe that $w \in C^\infty(\Omega)$ and that w has zero Cauchy values at $\partial\Omega$. Then w obviously solves (3.5) with $h = \nabla w$, and therefore $0 \neq h \in \mathcal{N}(\mathcal{A}^*)$.

For $h \in \overline{\mathcal{R}(\mathcal{A})}$ we can find the classical formulation of (3.3) using the fact that $h = \nabla v$ for some $v \in H^1(B)$ and $\Delta v = 0$. As a result we obtain the following transmission problem: find $w \in H^1_\diamond(B)$ such that

$$\begin{aligned} \Delta w &= 0 \text{ in } B \setminus \partial\Omega, \\ \partial_\nu w &= 0 \text{ on } \partial B, \\ \partial_\nu w|_- - \partial_\nu w|_+ &= v \cdot h \text{ on } \partial\Omega. \end{aligned} \tag{3.7}$$

Remark 3.1.3. The operator $\mathcal{A}^*\mathcal{A}$ is also very similar to the operator $L_\Omega L^*_\Omega$ in [26] that is used to construct so-called localized potentials. Both operators map from $L^2_\diamond(\partial B)$ to $L^2_\diamond(\partial B)$, and $\mathcal{A}^*\mathcal{A}f = w|_{\partial B}$, where w solves (3.3) with $h = \nabla u_0|_\Omega$, while for $L_\Omega L^*_\Omega f = \tilde{v}|_{\partial B}$, where $\tilde{v}$ solves

$$\iint_B \gamma \nabla \tilde{v} \cdot \nabla \psi \, dx = \iint_\Omega h \cdot \nabla \psi \, dx \quad \text{for all } \psi \in H^1_\diamond(B),$$

where $h = \nabla u|_\Omega$, and u solves (1.3). Hence the only difference is that for $\mathcal{A}^*\mathcal{A}$ we use the the background conductivity $\gamma_0 = 1$, while for $L_\Omega L^*_\Omega$ the actual conductivity γ is used.

In addition, the operator $\mathcal{T} : L^2(\Omega, \mathbb{R}^d) \to L^2(\Omega, \mathbb{R}^d)$ is defined by $h \mapsto q\,(h - \nabla w)$, where $w \in H^1_\diamond(B)$ solves

$$\iint_B (1 + q\chi_\Omega) \nabla w \cdot \nabla \psi \, dx = \iint_\Omega q\, h \cdot \nabla \psi \, dx \quad \text{for all } \psi \in H^1_\diamond(B). \tag{3.8}$$

Again, for $h \in \overline{\mathcal{R}(\mathcal{A})}$ we can formulate this as a transmission problem: find $w \in H^1_\diamond(B)$ such that

$$
\begin{aligned}
\Delta w &= 0 \text{ in } B \setminus \partial\Omega, \\
\partial_\nu w &= 0 \text{ on } \partial B, \\
(1+q)\,\partial_\nu w|_- - \partial_\nu w|_+ &= q\,(\nu \cdot h) \text{ on } \partial\Omega.
\end{aligned}
\tag{3.9}
$$

Throughout this chapter we restrict $\mathcal{T}$ to $\overline{\mathcal{R}(\mathcal{A})}$, and we observe that $\mathcal{T} : \overline{\mathcal{R}(\mathcal{A})} \to \overline{\mathcal{R}(\mathcal{A})}$ since q is constant and the potential w that solves the transmission problem (3.9) satisfies $\nabla w|_\Omega \in \overline{\mathcal{R}(\mathcal{A})}$. In addition, it is easy to show that $\mathcal{T}$ is self-adjoint.

By construction of the appearing operators the following factorization can be shown as it is done in [55].

Lemma 3.1.4. *Let $\mathcal{A}, \mathcal{T}$ be defined as above. Then:*

$$
\Lambda_0 - \Lambda = \mathcal{A}^* \mathcal{T} \mathcal{A}.
\tag{3.10}
$$

Proof. The weak formulation for the solution $u \in H^1_\diamond(B)$ of the direct problem for a given current pattern $f \in L^2_\diamond(\partial B)$ is

$$
\iint_B (1 + q\chi_\Omega)\nabla u \cdot \nabla \psi \, dx = \int_{\partial B} f\psi \, ds \quad \text{for all } \psi \in H^1_\diamond(B),
$$

while the weak formulation for the solution $u_0 \in H^1_\diamond(B)$ of the background direct problem is (3.1). By subtraction we obtain

$$
\iint_B (1 + q\chi_\Omega)\nabla(u_0 - u) \cdot \nabla\psi \, dx = \iint_\Omega q\nabla u_0 \cdot \nabla\psi \, dx \quad \text{for all } \psi \in H^1_\diamond(B).
$$

From this representation we observe that $(\Lambda_0 - \Lambda)f = \mathcal{G}\mathcal{A}f$, where the operator $\mathcal{G} : L^2(\Omega, \mathbb{R}^d) \to L^2_\diamond(\partial B)$ maps h to $w|_{\partial B}$, and $w \in H^1_\diamond(B)$ solves (3.8). This newly introduced operator $\mathcal{G}$ can now be decomposed as follows. We rewrite (3.8) to obtain

$$
\iint_B \nabla w \cdot \nabla\psi \, dx = \iint_\Omega q(h - \nabla w) \cdot \nabla\psi \, dx \quad \text{for all } \psi \in H^1_\diamond(B).
$$

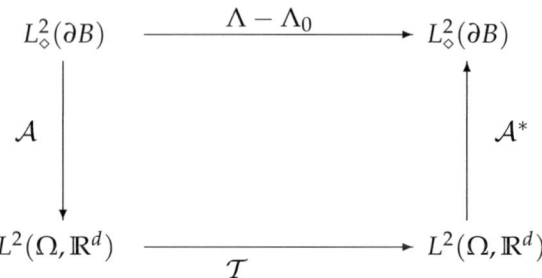

Figure 3.1: Diagram of operators and spaces for the factorization (3.10)

From this representation it is clear that the composition $\mathcal{G} = \mathcal{A}^*\mathcal{T}$ holds, which completes the proof.

$\square$

Figure 3.1 illustrates this new factorization schematically. Moreover, this factorization can be used to derive a new version of the Factorization method as it is done in [55]. This derivation is based on the coercivity of $\mathcal{T}$ (or of $-\mathcal{T}$, respectively) which follows immediately from Lemma 3.2.1. This coercivity admits the decomposition $|\mathcal{T}| = |\mathcal{T}|^{\frac{1}{2}}\,|\mathcal{T}|^{\frac{1}{2}}$, where $|\mathcal{T}|^{\frac{1}{2}} : \overline{\mathcal{R}(\mathcal{A})} \to \overline{\mathcal{R}(\mathcal{A})}$ is boundedly invertible. This result will also be used in Section 3.3.1.

The main idea of our method to determine the conductivity inside Ω is to make use of the factorization (3.10). The operators $\mathcal{A}, \mathcal{A}^*$ only contain information about the inclusion's boundary, hence they are known after the inclusion has been located. The information about the conductivity inside Ω lies inside the middle operator $\mathcal{T}$.

In the next section we take a closer look at $\mathcal{T}$ and especially its spectrum, while afterwards we show how the knowledge about this spectrum can be used to compute the conductivity contrast q.

3.2 The Spectrum of the Middle Operator

In this section we show some important properties of the operator $\mathcal{T}$ and its spectrum $\sigma(\mathcal{T})$. We start our considerations by a simple example in which basic properties of the spectrum $\sigma(\mathcal{T})$ can be observed.

3.2.1 A Radially Symmetric Example

We assume that B is the unit disc in $\mathbb{R}^2$ and the inclusion Ω is a concentric circle, i.e. $\Omega = B(0, R)$ with radius $R < 1$. The conductivity in B is then as follows:

$$\gamma(x) = \begin{cases} 1, & |x| > R, \\ 1 + q, & |x| < R, \end{cases}$$

and q is a constant such that $1 + q > 0$ and $q \neq 0$. Let $h \in \overline{\mathcal{R}(\mathcal{A})}$, i.e. there is $u \in H^1(\Omega)$ such that $h = \nabla u$ and $\Delta u = 0$ in Ω. Now the jump condition for the normal derivative on $\partial\Omega$ in the transmission problem (3.9) is as follows:

$$(1 + q)\, \partial_\nu w|_- - \partial_\nu w|_+ = q\, \partial_\nu u \text{ on } \partial\Omega. \tag{3.11}$$

By separation of variables we can represent u and w in polar coordinates:

$$u(r, \phi) = a_0 + \sum_{n=1}^{\infty} (a_n \cos(n\phi) + b_n \sin(n\phi))\, r^n, \ r < R,$$

$$w(r, \phi) = \begin{cases} c_0^- + \sum_{n=1}^{\infty} (c_n^- \cos(n\phi) + d_n^- \sin(n\phi))\, r^n, & r < R, \\ c_0^+ + \sum_{\substack{n=-\infty \\ n \neq 0}}^{\infty} (c_n^+ \cos(n\phi) + d_n^+ \sin(n\phi))\, r^n, & r > R. \end{cases}$$

The partial derivatives with respect to r are

$$\frac{\partial u}{\partial r}(r, \phi) = \sum_{n=1}^{\infty} n\, (a_n \cos(n\phi) + b_n \sin(n\phi))\, r^{n-1}, \ r < R,$$

$$\frac{\partial w}{\partial r}(r,\phi) = \begin{cases} \sum\limits_{n=1}^{\infty} n\left(c_n^- \cos(n\phi) + d_n^- \sin(n\phi)\right) r^{n-1}, & r < R, \\ \sum\limits_{\substack{n=-\infty \\ n \neq 0}}^{\infty} n\left(c_n^+ \cos(n\phi) + d_n^+ \sin(n\phi)\right) r^{n-1}, & r > R. \end{cases}$$

First we make use of the homogeneous Neumann boundary condition $\frac{\partial w}{\partial r} = 0$ at ∂B which implies that

$$\sum_{\substack{n=-\infty \\ n \neq 0}}^{\infty} n\left(c_n^+ \cos(n\phi) + d_n^+ \sin(n\phi)\right) = 0 \quad \text{for all } \phi \in [0, 2\pi).$$

From symmetry properties of the trigonometric functions we can deduce $c_{-n}^+ = c_n^+, d_{-n}^+ = -d_n^+$ $(n \in \mathbb{N})$ and thus

$$w(r,\phi) = \begin{cases} c_0^- + \sum\limits_{n=1}^{\infty} \left(c_n^- \cos(n\phi) + d_n^- \sin(n\phi)\right) r^n, & r < R, \\ c_0^+ + \sum\limits_{n=1}^{\infty} \left(c_n^+ \cos(n\phi) + d_n^+ \sin(n\phi)\right) (r^n + r^{-n}), & r > R. \end{cases}$$

Now we apply the continuity condition $w|_+ = w|_-$ at $\partial\Omega$ and obtain

$$c_0^- + \sum_{n=1}^{\infty} \left(c_n^- \cos(n\phi) + d_n^- \sin(n\phi)\right) R^n$$

$$= c_0^+ + \sum_{n=1}^{\infty} \left(c_n^+ \cos(n\phi) + d_n^+ \sin(n\phi)\right) (R^n + R^{-n}).$$

By comparison of the coefficients it follows that $c_0^- = c_0^+, c_n^+ = \frac{R^n}{R^n + R^{-n}} c_n^-$ and $d_n^+ = \frac{R^n}{R^n + R^{-n}} d_n^-$ $(n \in \mathbb{N})$. In order to keep our notations simple we denote $c_0 := c_0^-, c_n := c_n^-, d_n := d_n^-$ $(n \in \mathbb{N})$ and rewrite w as

$$w(r,\phi) = \begin{cases} c_0 + \sum\limits_{n=1}^{\infty} \left(c_n \cos(n\phi) + d_n \sin(n\phi)\right) r^n, & r < R, \\ c_0 + \sum\limits_{n=1}^{\infty} \frac{R^n}{R^n + R^{-n}} \left(c_n \cos(n\phi) + d_n \sin(n\phi)\right) (r^n + r^{-n}), \\ \qquad r > R. \end{cases}$$

The partial derivative of w with respect to r now is

$$\frac{\partial w}{\partial r}(r,\phi) = \begin{cases} \sum\limits_{n=1}^{\infty} n\,(c_n\cos(n\phi) + d_n\sin(n\phi))\,r^{n-1}, & r < R, \\ \sum\limits_{n=1}^{\infty} \frac{nR^n}{R^n+R^{-n}}\,(c_n\cos(n\phi) + d_n\sin(n\phi))\,(r^{n-1} - r^{-n-1}), \\ \qquad r > R, \end{cases}$$

and we can apply the transmission condition (3.11) which means that

$$(1+q) \sum_{n=1}^{\infty} n\,(c_n\cos(n\phi) + d_n\sin(n\phi))\,R^{n-1}$$
$$- \sum_{n=1}^{\infty} \frac{nR^n}{R^n + R^{-n}}\,(c_n\cos(n\phi) + d_n\sin(n\phi))\,(R^{n-1} - R^{-n-1})$$
$$= q \sum_{n=1}^{\infty} n\,(a_n\cos(n\phi) + b_n\sin(n\phi))\,R^{n-1}.$$

For the coefficients we deduce

$$(1+q)n\,c_n R^{n-1} - \frac{nR^n}{R^n + R^{-n}}c_n(R^{n-1} - R^{-n-1}) = q\,n\,a_n R^{n-1},$$
$$(1+q)n\,d_n R^{n-1} - \frac{nR^n}{R^n + R^{-n}}d_n(R^{n-1} - R^{-n-1}) = q\,n\,b_n R^{n-1},$$

and with a simple computation we obtain for $n \in \mathbb{N}$

$$c_n = \frac{q(R^{2n} + 1)}{2 + q(R^{2n} + 1)}a_n, \quad d_n = \frac{q(R^{2n} + 1)}{2 + q(R^{2n} + 1)}b_n.$$

Now we observe that the auxiliary operator $\tilde{T}$ that maps ∇u to ∇w has an eigensystem $\{\eta_n, \psi_n : n \in \mathbb{N}\}$ with eigenvalues

$$\eta_n = \frac{q(R^{2n} + 1)}{2 + q(R^{2n} + 1)}$$

converging to $\frac{q}{2+q}$ as n tends to infinity with corresponding eigenfunctions

$$\psi_n^{(1)}(r,\phi) = \cos(n\phi)r^n, \quad \psi_n^{(2)}(r,\phi) = \sin(n\phi)r^n.$$

Remembering that the operator $\mathcal{T}$ itself was defined as $\mathcal{T} = q(I - \tilde{\mathcal{T}})$ (compare page 64) we realize that $\mathcal{T}$ also has an eigensystem with the very same eigenfunctions $\psi_n^{(1)}, \psi_n^{(2)}$ ($n \in \mathbb{N}$). The eigenvalues are

$$\lambda_n = q(1 - \eta_n) = \frac{2q}{2 + q\,(R^{2n} + 1)},$$

and they converge towards $\lambda^* := \frac{2q}{2+q}$ as n tends to infinity.

During the next section we will show that also for arbitrary shapes of B and Ω the eigenvalues of $\mathcal{T}$ form a convergent sequence and that the spectrum of $\mathcal{T}$ consists only of this sequence of eigenvalues λ_n and its limit point λ^*. Furthermore, these considerations lead to the supposition that a representation $\mathcal{T} = \lambda^* I + K$ with a compact operator K holds and thus that $\mathcal{T}$ is a Fredholm operator. This Fredholm property will also be proven in the following section.

3.2.2 Constant Conductivity Contrast

Returning to the more general case, i.e. B and Ω satisfy the assumptions from Section 3.1, we assume again that the conductivity is such that $\gamma = 1 + q\chi_\Omega$, where $q \neq 0$ is real-valued constant and $1 + q > 0$. We start our considerations by giving bounds for $\sigma(\mathcal{T})$.

Lemma 3.2.1. *If $\lambda \in \sigma(\mathcal{T})$ then $\lambda \geq \frac{q}{1+q}$ and $\lambda < q$.*

Proof. Since $\mathcal{T}$ is a bounded self-adjoint operator, $\sigma(\mathcal{T})$ is a compact set in $\mathbb{R}$. Furthermore, it is a well-known fact that

$$\sigma(\mathcal{T}) \subset \left\{ \langle \mathcal{T}h, h \rangle : h \in L^2(\Omega, \mathbb{R}^d)\ \|h\| = 1 \right\}.$$

For the upper bound we use the weak formulation (3.8) and obtain for

$h \in L^2(\Omega, \mathbb{R}^d)$ with $\|h\| = 1$

$$\langle Th, h \rangle = \iint_\Omega q\,(h - \nabla w) \cdot h\,dx$$

$$= \iint_\Omega q\,|h|^2\,dx - \iint_\Omega q\,(\nabla w \cdot h)\,dx$$

$$= q - \iint_B (1 + q\chi_\Omega)\,|\nabla w|^2\,dx < q.$$

Equality would imply that $\nabla w \equiv 0$ in B from which we can easily conclude that $h = 0$, in contradiction to $\|h\| = 1$. Now it remains to show the lower bound where we again assume $\|h\| = 1$. Using (3.8) we obtain

$$\langle Th, h \rangle = \iint_\Omega q\,(h - \nabla w) \cdot h\,dx$$

$$= \iint_\Omega q\,|h - \nabla w|^2\,dx + \iint_\Omega q\,(h - \nabla w) \cdot \nabla w\,dx$$

$$= \iint_\Omega q\,|h - \nabla w|^2\,dx + \iint_B |\nabla w|^2\,dx$$

$$\geq \iint_\Omega \left(q\,|h|^2 - 2q\,(h \cdot \nabla w) + (1 + q)\,|\nabla w|^2 \right) dx$$

$$= \iint_\Omega \left(\left| \sqrt{1 + q}\,\nabla w - \frac{q}{\sqrt{1 + q}}h \right|^2 + \frac{q}{1 + q}\,|h|^2 \right) dx \geq \frac{q}{1 + q}.$$

$\square$

Remark 3.2.2. Note that the assertion of Lemma 3.2.1 holds for both $q > 0$ and $q < 0$ and in particular that $\frac{q}{1+q} < q$ holds independently of the sign of q. If $q > 0$, then T is coercive and $\sigma(T)$ lies on the positive real axis, while in the case $q < 0$ the operator $-T$ is coercive and $\sigma(T)$ is part of the negative real axis.

From now on we restrict T to $T : \overline{R(A)} \rightarrow \overline{R(A)}$. Soon we will see that T is closely related to certain boundary integral operators. We therefore proceed by investigating some properties of these boundary integral operators. For a detailed introduction into boundary integral equation methods we refer to [66], [19] or [20].

Let N be the Neumann function that is defined in (2.11) and let $SL : H_\diamond^{-\frac{1}{2}}(\partial\Omega) \to H^1(B)$ be the single layer potential. i.e.

$$SL\varphi(x) = \int_{\partial\Omega} N(x,y)\varphi(y)ds(y), \quad x \in B \setminus \partial\Omega.$$

In the following lemma we characterize the range of SL and show that on this range it is a boundedly invertible operator. In the corresponding proof the double layer potential $DL : H_\diamond^{\frac{1}{2}}(\partial B) \to H^1(B \setminus \partial\Omega)$,

$$DL\varphi(x) = \int_{\partial\Omega} \partial_{\nu(y)} N(x,y)\varphi(y)ds(y), \quad x \in B \setminus \partial\Omega$$

will also appear. The assumption $\Delta w = 0$ for $w \in H^1(B \setminus \partial\Omega)$ has to be understood in the sense of Remark 3.2.4.

Lemma 3.2.3. *Consider the space*

$$X := \left\{ w \in H_\diamond^1(B) : \Delta w = 0 \text{ in } B \setminus \partial\Omega, \, \partial_\nu w = 0 \text{ on } \partial B \right\},$$

equipped with the H^1-norm, then SL is a bijective operator from $H_\diamond^{-\frac{1}{2}}(\partial\Omega)$ onto X and the inverse $SL^{-1} : X \to H_\diamond^{-\frac{1}{2}}(\partial\Omega)$ is bounded.

Proof. Let $w = SL\varphi$ for $\varphi \in H_\diamond^{-\frac{1}{2}}(\partial\Omega)$, then $\Delta w = 0$ in $B \setminus \partial\Omega$. Using the properties of the Neumann function (see (2.11)) we obtain for $\varphi \in C_\diamond^2(\partial\Omega)$ that

$$\partial_\nu SL\varphi|_{\partial B} = \int_{\partial\Omega} \partial_\nu N(\cdot,y)\varphi(y) \, ds(y) \bigg|_{\partial B} = |\partial B|^{-1} \int_{\partial\Omega} \varphi \, ds = 0.$$

From the denseness of $C_\diamond^2(\partial\Omega)$ in $H_\diamond^{-\frac{1}{2}}(\partial\Omega)$ and continuity of the trace operator we obtain that $SL\varphi$ has zero Neumann values on ∂B for all $\varphi \in H_\diamond^{-\frac{1}{2}}(\partial B)$.

Furthermore, for $\varphi \in C^2_\diamond(\partial B)$ we obtain

$$\int\limits_{\partial B} SL\varphi \, ds = \int\limits_{\partial B} \int\limits_{\partial \Omega} N(x,y)\varphi(y) \, ds(y) \, ds(x)$$

$$= \int\limits_{\partial \Omega} \varphi(y) \int\limits_{\partial B} N(x,y) \, ds(x) \, ds(y) = 0,$$

and by the same denseness argument as before it follows that $\int\limits_{\partial B} SL\varphi \, ds = 0$ for all $\varphi \in H^{-\frac{1}{2}}_\diamond(\partial \Omega)$. We have thus shown that $\mathcal{R}(SL) \subset X$.

Assume that $SL\varphi = 0$ for some $\varphi \in H^{-\frac{1}{2}}_\diamond(\partial \Omega)$. Then the jump conditions on $\partial \Omega$ (see e.g. Theorem 6.11 in [66]) imply $\varphi = \partial_\nu SL\varphi|_- - \partial_\nu SL\varphi|_+ = 0$.

Now let $u \in X$, $x \in B \setminus \partial \Omega$, then we use the third Green identity in the form of Theorem 6.10 in [66] to show

$$SL\left(\partial_\nu u|_- - \partial_\nu u|_+\right)(x)$$
$$= SL\left(\partial_\nu u|_- - \partial_\nu u|_+\right)(x) - DL\left(u|_- - u|_+\right)(x)$$
$$= u(x),$$

and hence $u = SL\varphi$ with $\varphi = \partial_\nu u|_- - \partial_\nu u|_+ \in H^{-\frac{1}{2}}_\diamond(\partial \Omega)$. This shows that SL is an isomorphism between the spaces $H^{-\frac{1}{2}}_\diamond(\partial \Omega)$ and X.

In addition, X is a closed subspace of $H^1(B)$, which follows directly from the variational formulation for X (see Remark 3.2.4), and the open mapping theorem asserts that $SL^{-1} : X \to H^{-\frac{1}{2}}_\diamond(\partial \Omega)$ is bounded. $\quad\square$

Remark 3.2.4. A variational formulation for elements of the space X from Lemma 3.2.3 is given by

$$X = \left\{ w \in H^1_\diamond(B) : \iint\limits_\Omega \nabla w \cdot \nabla \psi \, dx = 0 \quad \forall \psi \in H^1_0(\Omega), \right.$$

$$\left. \iint\limits_{B \setminus \overline{\Omega}} \nabla w \cdot \nabla \psi \, dx = 0 \quad \forall \psi \in H^1(B \setminus \overline{\Omega}) : \psi|_{\partial \Omega} = 0 \right\}.$$

We proceed by considering boundary integral operators and define S : $C^2(\partial\Omega) \to L^2(\partial\Omega)$ by

$$S\varphi(x) = \int_{\partial\Omega} N(x,y)\varphi(y)ds(y), \quad x \in \partial\Omega.$$

Theorem 7.1 in [66] asserts that there is a bounded extension of S to $H^{-\frac{1}{2}}(\partial\Omega)$ and that the range of S is contained in $H^{\frac{1}{2}}(\partial\Omega)$. In addition, S is a self-adjoint operator (compare Chapter 7 of [66]).

Furthermore, we have to define $D : C^2(\partial\Omega) \to L^2(\partial\Omega)$ by

$$D\varphi(x) = \int_{\partial\Omega} \partial_{\nu_y} N(x,y)\varphi(y)ds(y), \quad x \in \partial\Omega.$$

Theorem 7.1 in [66] asserts that D has a bounded extension to $H^{\frac{1}{2}}(\partial\Omega)$ and that the range of D is contained in $H^{\frac{1}{2}}(\partial\Omega)$. Since we assumed the inclusion boundary $\partial\Omega$ to be of class C^2, we can apply Theorem 2.14 from [49] to show that D maps continuously from $H^{\frac{1}{2}}(\partial\Omega)$ into $H^1(\partial\Omega)$. Hence $D : H^{\frac{1}{2}}(\partial\Omega) \to H^{\frac{1}{2}}(\partial\Omega)$ is a compact operator.

Let us consider its adjoint $D^* : H^{-\frac{1}{2}}(\partial\Omega) \to H^{-\frac{1}{2}}(\partial\Omega)$ which is for $\varphi \in C^2(\partial\Omega)$ given by

$$D^*\varphi(x) = \int_{\partial\Omega} \partial_{\nu_x} N(x,y)\varphi(y)ds(y), \quad x \in \partial\Omega.$$

It can easily be shown that D^* is also well-defined as $D^* : H_\diamond^{-\frac{1}{2}}(\partial\Omega) \to H_\diamond^{-\frac{1}{2}}(\partial\Omega)$: let $\varphi \in H_\diamond^{-\frac{1}{2}}(\partial\Omega)$, then we have

$$\langle D^*\varphi, 1 \rangle_{H^{-\frac{1}{2}}(\partial\Omega) \times H^{\frac{1}{2}}(\partial\Omega)} = \langle \varphi, D1 \rangle_{H^{-\frac{1}{2}}(\partial\Omega) \times H^{\frac{1}{2}}(\partial\Omega)}.$$

From the jump relations (see Theorem 6.11 in [66]) we obtain $D1 = \frac{1}{2} + DL1|_-$. Using the second Green identity yields that $DL1|_- = 0$ which implies $D1 = const$ and thus that $\langle D^*\varphi, 1 \rangle = 0$.

Furthermore, since we assumed Ω to be a domain of class C^2, D^* is a compact operator, hence its spectrum consists at most of a sequence of eigenvalues and zero. In the following lemma we give bounds for the spectrum of D^*. The proof is based on the considerations in [59] where the derivations are carried out for the Dirichlet problem instead of the present Neumann problem. In addition, in [59] D^* is considered a map from the space of continuous functions on $\partial\Omega$ to itself, while we extend the assertion to $H_\diamond^{-\frac{1}{2}}(\partial\Omega)$.

Lemma 3.2.5. *All eigenvalues of $D^* : H_\diamond^{-\frac{1}{2}}(\partial\Omega) \to H_\diamond^{-\frac{1}{2}}(\partial\Omega)$ have absolute value less than or equal to $\frac{1}{2}$. Furthermore, $\frac{1}{2}$ is not an eigenvalue of D^* but $-\frac{1}{2}$ is.*

Proof. Define $(\cdot, \cdot) : H_\diamond^{-\frac{1}{2}}(\partial\Omega) \times H_\diamond^{-\frac{1}{2}}(\partial\Omega) \to \mathbb{R}$ by

$$(\varphi, \psi) = \langle \varphi, S\psi \rangle_{H^{-\frac{1}{2}}(\partial\Omega) \times H^{\frac{1}{2}}(\partial\Omega)}.$$

$(\cdot, \cdot)$ is a well-defined bilinear form, since the boundary integral operator S maps from $H_\diamond^{-\frac{1}{2}}(\partial\Omega)$ to $H^{\frac{1}{2}}(\partial\Omega)$. It is symmetric since S is self-adjoint.

Furthermore, we show that $(\cdot, \cdot)$ is positive definite on $H_\diamond^{-\frac{1}{2}}(\partial\Omega)$ using the first Green identity:

$$(\varphi, \varphi) = \langle \varphi, S\varphi \rangle = \langle \partial_\nu(SL\varphi)|_- - \partial_\nu(SL\varphi)|_+ , S\varphi \rangle$$
$$= \iint_\Omega \nabla(SL\varphi) \cdot \nabla(SL\varphi)\, dx + \iint_{B\setminus\overline{\Omega}} \nabla(SL\varphi) \cdot \nabla(SL\varphi)\, dx \geq 0,$$

and from this representation it is easy to see that $(\varphi, \varphi) = 0$ implies $\nabla SL\varphi = 0$ in $B \setminus \partial\Omega$ and thus $\varphi = 0$.

We now show that D^* is symmetric with respect to this newly defined scalar product:

$$(D^*\varphi, \psi) = \frac{1}{2} \langle \partial_\nu(SL\varphi)|_- + \partial_\nu(SL\varphi)|_+ , S\psi \rangle$$
$$= \frac{1}{2} \iint_\Omega \nabla(SL\varphi) \cdot \nabla(SL\psi)\, dx - \frac{1}{2} \iint_{B\setminus\overline{\Omega}} \nabla(SL\varphi) \cdot \nabla(SL\psi)\, dx.$$

This expression is symmetric with respect to φ and ψ, hence $(D^*\varphi, \psi) = (\varphi, D^*\psi)$. Altogether we have shown that $\sigma(D^*)$ is a compact subset of $\mathbb{R}$.

Now define $R, \hat{R} : X \to \mathbb{R}_{\geq 0}$ by

$$R(u) = \iint_{\Omega} |\nabla u|^2 \, dx, \qquad \hat{R}(u) = \iint_{B \setminus \overline{\Omega}} |\nabla u|^2 \, dx,$$

where the space X is defined as in Lemma 3.2.3. Using the considerations from above we now derive a connection between $\sigma(D^*)$ and $R, \hat{R}$:

$$\sup\{\sigma(D^*)\} = \sup \left\{ \frac{(D^*\varphi, \varphi)}{(\varphi, \varphi)} : \varphi \in H_\diamond^{-\frac{1}{2}}(\partial\Omega), \, \varphi \neq 0 \right\}$$

$$= \sup \left\{ \frac{1}{2} \frac{R(SL\varphi) - \hat{R}(SL\varphi)}{R(SL\varphi) + \hat{R}(SL\varphi)} : \varphi \in H_\diamond^{-\frac{1}{2}}(\partial\Omega), \, \varphi \neq 0 \right\}$$

Since the range of SL coincides with X by Lemma 3.2.3, this implies that

$$\sup\{\sigma(D^*)\} = \sup \left\{ \frac{1}{2} \frac{R(u) - \hat{R}(u)}{R(u) + \hat{R}(u)} : u \in X, \, u \neq 0 \right\},$$

and in the same way it can be shown that

$$\inf\{\sigma(D^*)\} = \inf \left\{ \frac{1}{2} \frac{R(u) - \hat{R}(u)}{R(u) + \hat{R}(u)} : u \in X, \, u \neq 0 \right\}.$$

From these bounds we conclude that $\sigma(D^*) \subset \left[-\frac{1}{2}, \frac{1}{2} \right]$. Assume that $\mu = \frac{1}{2}$ is an eigenvalue of D^*. Then there is $u \in X$, $u \neq 0$ such that $\hat{R}(u) = 0$, which implies that $\nabla u \equiv 0$ in $B \setminus \overline{\Omega}$. We know that $u \in H_\diamond^1(B)$, hence $u \equiv 0$ in $B \setminus \overline{\Omega}$, and since u is continuous across $\partial\Omega$ we deduce that $u|_- = 0$. Uniqueness of the Dirichlet problem for the Laplace equation in Ω now implies $u \equiv 0$ in Ω, a contradiction to the assumption.

Let $u \in H^1(B)$ be such that $u \equiv 1$ in Ω and $\partial_\nu u = 0$ ob ∂B. Such an u exists if there is a weak solution $\tilde{u} \in H^1(B \setminus \overline{\Omega})$ to $\Delta\tilde{u} = 0$ in $B \setminus \overline{\Omega}$, $\tilde{u}|_{\partial\Omega} = 1$, $\partial_\nu \tilde{u}|_{\partial B} = 0$. The corresponding homogeneous problem is almost the same but with a zero Dirichlet boundary condition at $\partial\Omega$. Hence its weak formulation is as follows: find $\tilde{u}_0 \in H^1(B \setminus \overline{\Omega})$ that solves

$$\iint_{B \setminus \overline{\Omega}} \nabla \tilde{u}_0 \cdot \nabla \psi = 0 \quad \text{for all } \psi \in H^1(B \setminus \overline{\Omega}) : \psi|_{\partial\Omega} = 0 \qquad (3.12)$$

and $\tilde{u}_0|_{\partial\Omega} = 0$. Choosing $\psi = \tilde{u}_0$ in (3.12) yields $\tilde{u}_0 = const$ in $B \setminus \overline{\Omega}$, and the homogeneous Dirichlet boundary condition implies $\tilde{u}_0 = 0$. This shows that the homogeneous mixed boundary value problem (3.12) has only the trivial solution, and Theorem 4.10 in [66] asserts that there is a unique $\tilde{u}$ as required and thus that there is $u \in H^1(B)$ as stated above. Now we define $\hat{u} \in H^1(B)$ by $\hat{u} = u + C$, where the constant C is chosen such that $\hat{u}|_{\partial B} \in H_\diamond^{\frac{1}{2}}(\partial B)$ and observe that $\hat{u}|_\Omega \equiv 1 + C$ and that $\hat{u} \in X$.

Now for $\hat{u}$ we deduce that $\frac{R(\hat{u}) - \hat{R}(\hat{u})}{R(\hat{u}) + \hat{R}(\hat{u})} = -1$, and using Lemma 3.2.3 yields that there is $\varphi \in H_\diamond^{-\frac{1}{2}}(\partial\Omega)$ such that $SL\varphi = \hat{u}$. In addition, φ is an eigenfunction of D^* for the eigenvalue $\mu = -\frac{1}{2}$, which follows immediately from the jump conditions for the single layer potential.

$\square$

We proceed by deriving a connection between $\mathcal{T}$ and D^* in order to characterize the spectrum of $\mathcal{T}$. In Lemma 3.2.1 we have already shown that $\sigma(\mathcal{T}) \subset \left[\frac{q}{1+q}, q\right)$ which allows us to restrict our considerations to this interval.

Theorem 3.2.6. λ *is an eigenvalue of* $\mathcal{T}$ *if and only if* $\mu = -\frac{2+q}{2q} + \frac{1}{\lambda}$ *is an eigenvalue of* D^* *and* $\mu \neq -\frac{1}{2}$.

Proof. "$\Longrightarrow$": Let $\lambda \in \left[\frac{q}{1+q}, q\right)$ be an eigenvalue of $\mathcal{T}$, i.e. there exists $h \in \overline{\mathcal{R}(\mathcal{A})}$, $h \neq 0$ such that $\mathcal{T}h = \lambda h$ and $h = \nabla u$ for some $u \in H^1(\Omega)$ satisfying $\Delta u = 0$ in Ω. From the definition of $\mathcal{T}$ we obtain

$$q(\nabla u - \nabla w) = \lambda \nabla u \quad \text{in } \Omega, \tag{3.13}$$

where w is the weak solution of the transmission boundary value problem (3.9). From (3.13) we obtain $(q - \lambda)\nabla u = q\nabla w$ and thus $u = \frac{q}{q-\lambda}w + c$ with some constant c ($\lambda \neq q$ by assumption). The transmission condition in (3.9) can now be written in terms of w:

$$\underbrace{\left(1 + q - \frac{q^2}{(q - \lambda)}\right)}_{=:\rho} \partial_\nu w|_- - \partial_\nu w|_+ = 0 \quad \text{on } \partial\Omega. \tag{3.14}$$

Since $w \in X$, w can be represented by a single layer potential $w = SL\varphi$ with a density $\varphi \in H_\diamond^{-\frac{1}{2}}(\partial\Omega)$. Using the jump relations for the single layer potential and (3.14), leads to the the integral equation

$$\rho\left(\frac{1}{2}\varphi + D^*\varphi\right) - \left(-\frac{1}{2}\varphi + D^*\varphi\right) = 0$$

on $\partial\Omega$ which can be simplified to

$$\frac{1}{2}\frac{\rho+1}{\rho-1}\varphi + D^*\varphi = 0, \tag{3.15}$$

since $\rho = 1$ would imply $\lambda = 0$, a contradiction to $\sigma(\mathcal{T}) \subset \left[\frac{q}{1+q}, q\right)$. Equation (3.15) has a nontrivial solution if and only if $\mu := -\frac{1}{2}\frac{\rho+1}{\rho-1}$ is an eigenvalue of D^*. It remains to show that $-\frac{1}{2}\frac{\rho+1}{\rho-1} = -\frac{2+q}{2q} + \frac{1}{\lambda}$ which is a simple computation. In addition, $\mu \neq -\frac{1}{2}$ since in this case we have $\lambda = q$, another contradiction to $\sigma(\mathcal{T}) \subset \left[\frac{q}{1+q}, q\right)$.

"$\Longleftarrow$": Now let $\mu \in \left(-\frac{1}{2}, \frac{1}{2}\right)$ be an eigenvalue of D^*. Then the integral equation

$$-\mu\varphi + D^*\varphi = 0$$

has a nontrivial solution $\varphi \in H_\diamond^{-\frac{1}{2}}(\partial\Omega)$, and the single layer ansatz $w = SL\varphi$ provides $w \in H_\diamond^1(B)$ such that $\Delta w = 0$ in $B \setminus \partial\Omega$, and $\partial_\nu w = 0$ on ∂B. Defining $\rho := \frac{2\mu-1}{2\mu+1}$, we realize that (3.14) holds. Define $\lambda := \frac{q(1-\rho)}{(1-\rho)+q}$ and $u := \frac{q}{q-\lambda}w + c$ with a constant c, then w solves (3.9) with ∇u instead of h, and we obtain $q(\nabla u - \nabla w) = \lambda\nabla u$ in Ω. Since $\mu \neq -\frac{1}{2}$ we know from the jump relations that $\partial_\nu w|_- \neq 0$ and thus w, u are non-constant in Ω.

$\square$

Remark 3.2.7. In our proof we had to exclude $\mu = -\frac{1}{2}$ which corresponds to $\lambda = q$, the upper bound for $\sigma(\mathcal{T})$ from Lemma 3.2.1 that is not attained by $\sigma(\mathcal{T})$. Since $\frac{1}{2}$ is not an eigenvalue of D^* we now even know from the correspondence $\mu = -\frac{2+q}{2q} + \frac{1}{\lambda}$ that $\frac{q}{1+q}$ is no eigenvalue of $\mathcal{T}$. This can also be observed in the example from Section 3.2.1 where the eigenvalues are $\lambda_n = \frac{2q}{2+q(R^{2n}+1)}$ $(n \in \mathbb{N})$. Since $n = 0$ is excluded, there is no eigenvalue equal to $\frac{q}{q+1}$.

In the following corollary and theorem we use this correspondence between T and the integral operator D^* and similar techniques as in the proof of Theorem 3.2.6 to obtain further results about the structure of $\sigma(T)$. Afterwards, in Theorem 3.2.10, we will show that T is a Fredholm operator which yields, together with the spectral theorem for compact, self-adjoint operators, an explicit characterization of $\sigma(T)$ which is a stronger result than the following two assertions.

Corollary 3.2.8. *If there is an infinite sequence of eigenvalues λ_n of T, then they converge towards $\lambda^* := \frac{2q}{2+q}$.*

Proof. Let $(\lambda_n)_{n \in \mathbb{N}}$ be a sequence of eigenvalues of T. Define $\mu_n := -\frac{2+q}{2q} + \frac{1}{\lambda_n}$ $(n \in \mathbb{N})$. Then Theorem 3.2.6 states that $(\mu_n)_{n \in \mathbb{N}}$ is a sequence of eigenvalues of D^*. Since D^* is compact, the only possible accumulation point of its eigenvalues is zero, which implies that $\lambda_n \to \frac{2q}{2+q}$ $(n \to \infty)$. $\square$

Theorem 3.2.9. *If $\lambda \in \sigma(T)$, then λ is an eigenvalue of T or $\lambda = \lambda^*$, where $\lambda^* = \frac{2q}{2+q}$.*

Proof. Assume that $\lambda \in \left(\frac{q}{1+q}, q\right)$ with $\lambda \neq \lambda^*$ is not an eigenvalue of T. Then $T - \lambda I : \overline{\mathcal{R}(A)} \to \overline{\mathcal{R}(A)}$ is one-to-one. If we can show that $T - \lambda I$ is also onto, then we know by the open mapping theorem that it has a bounded inverse and thus that $\lambda \notin \sigma(T)$.

In order to show surjectivity we assume that $g \in \overline{\mathcal{R}(A)}$, i.e. $g = \nabla v$ for some $v \in H^1(\Omega)$ such that $\Delta v = 0$ in Ω, and we construct $h \in \overline{\mathcal{R}(A)}$ satisfying $(T - \lambda I)h = g$. To this end we first consider an equivalent boundary integral equation: find $\varphi \in H_\diamond^{-\frac{1}{2}}(\partial\Omega)$ that solves

$$-\mu\varphi + D^*\varphi = -\frac{1}{\lambda}\,\partial_\nu v|_- \tag{3.16}$$

with $\mu := -\frac{2+q}{2q} + \frac{1}{\lambda}$. Since λ is neither an eigenvalue of $\mathcal{T}$ nor equal to λ^* we deduce from Theorem 3.2.6 that $\mu \neq 0$ and μ is not an eigenvalue of D^* and thus that (3.16) has a unique solution $\varphi \in H_\diamond^{-\frac{1}{2}}(\partial\Omega)$. Using this solution we now define $w \in X$ by $w := SL\varphi$, where X denotes the space defined in Lemma 3.2.3. Using the jump conditions for the single layer potential we observe after a short computation that w satisfies the following jump condition:

$$\left(1 + q - \frac{q^2}{q-\lambda}\right) \partial_\nu w|_- - \partial_\nu w|_+ = \frac{q}{q-\lambda} \partial_\nu v|_- ,$$

which can be converted to

$$(1+q)\,\partial_\nu w|_- - \partial_\nu w|_+ = \frac{q^2}{q-\lambda} \partial_\nu w|_- + \frac{q}{q-\lambda} \partial_\nu v|_- . \tag{3.17}$$

Now we define $h \in \overline{\mathcal{R}(\mathcal{A})}$ by $h := \frac{1}{q-\lambda}\nabla v + \frac{q}{q-\lambda}\nabla w$, and by comparison of (3.17) with the jump condition in (3.9) we obtain that $\mathcal{T}h = q(h - \nabla w)$ and thus that

$$(\mathcal{T} - \lambda I)\,h = (q - \lambda)h - q\nabla w = \nabla v + q\nabla w - q\nabla w = \nabla v.$$

This shows that $(\mathcal{T} - \lambda I)h = g$, and the proof is complete. $\qquad\square$

The next theorem states that $\mathcal{T}$ is a Fredholm operator with index zero.

Theorem 3.2.10. *Let* $\lambda^* = \frac{2q}{2+q}$, *then the operator* $K : \overline{\mathcal{R}(\mathcal{A})} \to \overline{\mathcal{R}(\mathcal{A})}$ *defined by* $K := \mathcal{T} - \lambda^* I$ *is compact.*

Proof. Let $\nabla u \in \overline{\mathcal{R}(\mathcal{A})}$, then $K\nabla u = q\,(\nabla u - \nabla w) - \frac{2q}{2+q}\nabla u = \frac{q^2}{2+q}\nabla u - q\nabla w$, and $w \in H_\diamond^1(B)$ is the unique solution to (3.8) for $h = \nabla u$. Since q is constant we can define $v \in H^1(\Omega)$ with $\nabla v \in \overline{\mathcal{R}(\mathcal{A})}$ by $v := \frac{q^2}{2+q}u - qw$.

Our aim is to decompose the map $K : \nabla u \mapsto \nabla v$ into several bounded operators out of which one is compact and start with (3.8):

$$\iint_B (1 + q\chi_\Omega)\nabla w \cdot \nabla\psi\,dx = \iint_\Omega q\nabla u \cdot \nabla\psi\,dx \quad \text{for all } \psi \in H_\diamond^1(B),$$

and obtain using $q\nabla u = \frac{2+q}{q}\nabla v + (2+q)\nabla w$ that

$$\iint_{B\backslash\overline{\Omega}} \nabla w \cdot \nabla\psi\,dx + \iint_{\Omega} \nabla w \cdot \nabla\psi\,dx + \iint_{\Omega} q\nabla w \cdot \nabla\psi\,dx$$

$$= \iint_{\Omega} \frac{2+q}{q}\nabla v \cdot \nabla\psi\,dx + \iint_{\Omega} (2+q)\nabla w \cdot \nabla\psi\,dx,$$

which can be transformed to

$$\iint_{B\backslash\overline{\Omega}} \nabla w \cdot \nabla\psi\,dx - \iint_{\Omega} \nabla w \cdot \nabla\psi\,dx = \iint_{\Omega} \frac{2+q}{q}\nabla v \cdot \nabla\psi\,dx \qquad (3.18)$$

for all $\psi \in H_\diamond^1(B)$. This weak formulation formulation for w and some given v with $\nabla v \in \overline{\mathcal{R}(A)}$ corresponds to the transmission problem to find $w \in X$ where the space X is defined in Lemma 3.2.3 and

$$-\partial_\nu w|_+ - \partial_\nu w|_- = \frac{2+q}{q}\,\partial_\nu v|_- \quad \text{on } \partial\Omega.$$

As already done in the preceding proofs, we represent w by a single layer potential $w = SL\varphi$ with a density $\varphi \in H_\diamond^{-\frac{1}{2}}(\partial\Omega)$ that solves the integral equation

$$-\left(\frac{1}{2}\varphi + D^*\varphi\right) - \left(-\frac{1}{2}\varphi + D^*\varphi\right) = \frac{2+q}{q}\,\partial_\nu v|_-,$$

which can be simplified to

$$-2D^*\varphi = \frac{2+q}{q}\,\partial_\nu v|_-.$$

Now we return to the operator $K : \overline{\mathcal{R}(A)} \rightarrow \overline{\mathcal{R}(A)}$, $\nabla u \mapsto \nabla v$ and show that it is a composition of several bounded operators out of which at least one is compact. Let the space $X \subset H^1(B)$ be defined as in Lemma 3.2.3. In addition, let the auxiliary operator $\tilde{T} : \overline{\mathcal{R}(A)} \rightarrow X$ be defined by $\nabla u \mapsto w$, then the considered composition of operators is as follows:

$$K : \nabla u \xrightarrow{\tilde{T}} w \xrightarrow{SL^{-1}} \varphi \xrightarrow{D^*} -\frac{2+q}{2q} \partial_\nu v|_- \longmapsto \nabla v.$$

The map $\tilde{T}$ is a bounded operator. The single layer potential $SL :$ $H_\diamond^{-\frac{1}{2}}(\partial\Omega) \to X$ is bounded, bijective and has a bounded inverse $SL^{-1} :$ $X \to H_\diamond^{-\frac{1}{2}}(\partial\Omega)$ (see Lemma 3.2.3), thus the map $w \mapsto \varphi$ is also bounded. Since $\partial\Omega$ is of class C^2, $D^* : H_\diamond^{-\frac{1}{2}}(\partial\Omega) \to H_\diamond^{-\frac{1}{2}}(\partial\Omega)$ is a compact operator. The operator that maps the Neumann boundary value $\frac{2+q}{2q}\partial_\nu v|_- \in$ $H_\diamond^{-\frac{1}{2}}(\partial\Omega)$ to the corresponding weak solution $v \in H_\diamond^1(\Omega)$ of the Laplace equation in Ω as well as the weak derivative $v \mapsto \nabla v \in \mathcal{R}(\mathcal{A})$ are again bounded, which completes the proof. $\square$

The last conclusions in the proof show in particular that K is one-to-one because it is a composition of injective operators. Since $\mathcal{T}$ is self-adjoint we conclude that K is also self-adjoint. The spectral theorem for compact, self-adjoint operators applied to K yields the following corollary.

Corollary 3.2.11. *There is an orthonormal system of eigenfunctions of $\mathcal{T}$ in* $\overline{\mathcal{R}(\mathcal{A})}$ *with corresponding eigenvalues λ_n ($n \in \mathbb{N}$) converging towards λ^*. The spectrum $\sigma(\mathcal{T})$ consists of these eigenvalues λ_n ($n \in \mathbb{N}$) and their limit point λ^*.*

As already mentioned above, this result also implies the assertions from Corollary 3.2.8 and Theorem 3.2.9.

3.2.3 Several Conductivity Contrasts

In the previous section we assumed that there is only one inclusion having a constant conductivity. The next further development of our considerations is the case in which there are N inclusions. This problem setting is formulated in the following assumption.

Assumption 3.2.12. Let $\Omega_1, \ldots, \Omega_N$ be N separated C^2-domains in B, i.e. $\overline{\Omega}_i \cap \overline{\Omega}_j = \emptyset$ ($i, j = 1, \ldots, N$, $i \neq j$). Let the conductivity distribution γ satisfy

$$\gamma(x) = \begin{cases} 1 + q_j, & x \in \Omega_j \quad (j = 1, \ldots, N), \\ 1, & \text{otherwise.} \end{cases}$$

The constants q_j are such that $1 + q_j > 0$ for all $j = 1, \ldots, N$. By Ω we denote the union $\Omega = \Omega_1 \cup \cdots \cup \Omega_N$, while we denote $\gamma - 1|_\Omega$ by q. We also assume without loss of generality that the conductivity contrasts are mutually different from each other since for $q_j = q_k$ we can subsume Ω_j and Ω_k under one inclusion. As before, $B \setminus \overline{\Omega}$ is assumed to be connected.

The operators of our factorization of $\Lambda_0 - \Lambda$ now have the following mapping properties:

$$\mathcal{A} : L^2_\diamond(\partial B) \to L^2(\Omega_1, \mathbb{R}^2) \times \cdots \times L^2(\Omega_N, \mathbb{R}^2)$$
$$f \mapsto \left(\nabla u_0|_{\Omega_1}, \ldots, \nabla u_0|_{\Omega_N} \right)^\top,$$

where $u_0 \in H^1_\diamond(B)$ solves the background direct problem (3.2). For the adjoint $\mathcal{A}^*$ we obtain

$$\mathcal{A}^* : L^2(\Omega_1, \mathbb{R}^2) \times \cdots \times L^2(\Omega_N, \mathbb{R}^2) \to L^2_\diamond(\partial B),$$
$$(h_1, \ldots, h_N)^\top \mapsto v|_{\partial B},$$

and $v \in H^1_\diamond(B)$ is the solution to

$$\iint_B \nabla v \cdot \nabla \psi \, dx = \sum_{j=1}^N \iint_{\Omega_j} h_j \cdot \nabla \psi \, dx \quad \text{for all } \psi \in H^1_\diamond(B).$$

For the middle operator we obtain

$$\mathcal{T} : L^2(\Omega_1, \mathbb{R}^2) \times \cdots \times L^2(\Omega_N, \mathbb{R}^2) \to L^2(\Omega_1, \mathbb{R}^2) \times \cdots \times L^2(\Omega_N, \mathbb{R}^2)$$
$$(h_1, \ldots, h_N)^\top \mapsto \left(q_1 \left(h_1 - \nabla w|_{\Omega_1} \right), \ldots, q_N \left(h_N - \nabla w|_{\Omega_N} \right) \right)^\top,$$

where $w \in H^1_\diamond(B)$ solves

$$\iint_B \nabla w \cdot \nabla \psi \, dx + \sum_{j=1}^N \iint_{\Omega_j} q_j \nabla w \cdot \nabla \psi \, dx = \sum_{j=1}^N \iint_{\Omega_j} q_j h_j \cdot \nabla \psi \, dx \quad (3.19)$$

for all $\psi \in H^1_\diamond(B)$. As in the previous section we start our considerations by giving bounds for $\sigma(\mathcal{T})$ and adapt Lemma 3.2.1 to the present case.

Lemma 3.2.13. *If $\lambda \in \sigma(\mathcal{T})$, then $\lambda \geq \frac{q_{min}}{1+q_{min}}$ and $\lambda < q_{max}$, where $q_{min} = \min\{q_1, \ldots, q_N\}$ and $q_{max} = \max\{q_1, \ldots, q_N\}$.*

Proof. Analogously to the proof of Lemma 3.2.1 we show for $\lambda \in \sigma(\mathcal{T})$ that $\lambda < q_{max}$ and $\lambda \geq \min\left\{\frac{q_j}{1+q_j} : j = 1, \ldots, N\right\}$. It remains to show that $\frac{q_j}{1+q_j} \geq \frac{q_{min}}{1+q_{min}}$ for $j = 1, \ldots, N$, which can be obtained from the equality $\frac{q_j}{1+q_j} = 1 - \frac{1}{1+q_j}$. $\qquad\square$

Remark 3.2.14. If we assume that all the conductivity contrasts q_j ($j = 1, \ldots, N$) are positive, then Lemma 3.2.13 yields injectivity of the operator $\mathcal{T}$. The same holds if $q_1, \ldots, q_N$ are negative, respectively. However, $\mathcal{T}$ is injective even if the contrasts q_j ($j = 1, \ldots, N$) have different signs. This assertion correspronds to Lemma 2.1.5 but for the new factorization (3.10) instead of (2.5). However, the proof for the new factorization (3.10) is more elementary:

Assume that there is $h \in L^2(\Omega, \mathbb{R}^d)$ such that $\mathcal{T}h = 0$, then, since none of the q_j is zero, we have that $\nabla w = h$ in the whole of Ω. Using this equality, (3.19) reduces to

$$\iint_B \nabla w \cdot \nabla \psi \, dx = 0 \quad \text{for all } \psi \in H^1_\diamond(B).$$

Setting $\psi = w$ and yields that $w = const$ in B and thus $h = 0$ in Ω.

It is easy to check that the assertions of Lemma 3.1.1 hold here as well, and in particular that $\overline{\mathcal{R}(\mathcal{A})}$ has the representation (3.6).

Furthermore, we now restrict $\mathcal{T}$ to $\mathcal{T} : \overline{\mathcal{R}(\mathcal{A})} \rightarrow \overline{\mathcal{R}(\mathcal{A})}$ as we did in Section 3.2.2 in order to derive further results about $\sigma(\mathcal{T})$ in the case of N different conductivity contrasts.

We start by formulating an assertion corresponding to Theorem 3.2.6 for which we have to introduce some notations for the appearing boundary integral operators. The single layer potential $SL_j : H_\diamond^{-\frac{1}{2}}(\partial\Omega_j) \rightarrow H_\diamond^1(B)$ that corresponds to the particular inclusion boundary $\partial\Omega_j$ is for $j = 1, \ldots, N$ defined as the usual single layer potential SL but restricted to $\partial\Omega_j$:

$$SL_j\varphi(x) = \int_{\partial\Omega_j} N(x,y)\varphi(y)\,ds(y) \quad (x \in B \setminus \partial\Omega_j).$$

The boundary integral operators $D_j^* : H_\diamond^{-\frac{1}{2}}(\partial\Omega_j) \rightarrow H_\diamond^{\frac{1}{2}}(\partial\Omega_j)$ $(j = 1,\dots,N)$ are for densities $\varphi \in C_\diamond^2(\partial\Omega_j)$ defined by

$$D_j^* \varphi_j(x) = \int_{\partial\Omega_j} \partial_{\nu(x)} N(x,y)\varphi_j(y)\,ds(y) \quad (x \in \partial\Omega_j),$$

and as in Section 3.2.2 we use their bounded extensions to $H_\diamond^{-\frac{1}{2}}(\partial\Omega_j)$.

The operators $D_{j\to k}^*$ $(j,k = 1,\dots,N, j \neq k)$ are defined by the same integral, but evaluated on $\partial\Omega_k$ instead of $\partial\Omega_j$. This means in particular that $D_{j\to k}^*$ is the normal derivative of SL_j at the boundary $\partial\Omega_k$.

Theorem 3.2.15. *Let $\Omega_1,\dots,\Omega_N$ be the inclusions with corresponding conductivity contrasts $q_1,\dots,q_N$. Then $\lambda \in \mathbb{R}$ is an eigenvalue of $\mathcal{T}$ if and only if there are $\mu_1,\dots,\mu_N \in \mathbb{R}$ such that there is a solution $0 \neq (\varphi_1,\dots,\varphi_N)^\top \in H_\diamond^{-\frac{1}{2}}(\partial\Omega_1) \times \cdots \times H_\diamond^{-\frac{1}{2}}(\partial\Omega_N)$ to the system of integral equations*

$$\left(-\mu_j I + D_j^*\right)\varphi_j + \sum_{\substack{k=1 \\ k\neq j}}^{N} D_{k\to j}^* \varphi_k = 0, \quad (j=1,\dots,N), \tag{3.20}$$

such that the term $\frac{2q_j}{2\mu_j q_j + 2 + q_j}$ is identical for all $j = 1,\dots,N$. In this case $\lambda = \frac{2q_j}{2\mu_j q_j + 2 + q_j}$ $(j = 1,\dots,N)$.

Proof. "$\Longrightarrow$": Let λ be an eigenvalue of $\mathcal{T}$, i.e. there is $\nabla u \in \overline{\mathcal{R}(\mathcal{A})}$ such that $\mathcal{T}\nabla u|_{\Omega_j} = q_j\,(\nabla u - \nabla w)|_{\Omega_j} = \lambda\,\nabla u|_{\Omega_j}$ for all $j = 1,\dots,N$, and w solves (3.19) with $h = \nabla u$. We distinguish between two different cases.

Case 1: $\lambda \neq q_j$ for all $j = 1,\dots,N$. In this case we proceed as in the proof of Theorem 3.2.6 and express u in terms of w by $u|_{\Omega_j} = \frac{q_j}{q_j - \lambda}\,w|_{\Omega_j} +$

c_j $(j = 1, \ldots, N)$ and obtain the transmission conditions

$$\underbrace{\left(1 + q_j - \frac{q_j^2}{q_j - \lambda} \right)}_{=: \rho_j} \partial_\nu w|_- - \partial_\nu w|_+ = 0 \quad \text{on } \partial \Omega_j \ (j = 1, \ldots, N). \quad (3.21)$$

Hence w can be represented by a sum of single layer potentials $w = \sum_{k=1}^{N} SL_k \varphi_k$ with densities $\varphi_k \in H_\circ^{-\frac{1}{2}}(\partial \Omega_k)$ $(k = 1, \ldots, N)$. Using the jump relations for the single layer potential the normal derivatives of w on $\partial \Omega_j$ are as follows:

$$\partial_\nu w|_- = \frac{1}{2} \varphi_j + D_j^* \varphi_j + \sum_{\substack{k=1 \\ k \neq j}}^{N} D_{k \to j}^* \varphi_k \quad \text{on } \partial \Omega_j,$$

$$\partial_\nu w|_+ = -\frac{1}{2} \varphi_j + D_j^* \varphi_j + \sum_{\substack{k=1 \\ k \neq j}}^{N} D_{k \to j}^* \varphi_k \quad \text{on } \partial \Omega_j.$$

The transmission condition (3.21) implies that the densities $\varphi_1, \ldots, \varphi_N$ solve the system of integral equations

$$\underbrace{\frac{1}{2} \frac{\rho_j + 1}{\rho_j - 1}}_{=: -\mu_j} \varphi_j + D_j^* \varphi_j + \sum_{\substack{k=1 \\ k \neq j}}^{N} D_{k \to j}^* \varphi_k = 0 \quad \text{on } \partial \Omega_j \ (j = 1, \ldots, N),$$

since $\rho_j = 1$ would imply $\lambda = 0$, a contradiction to injectivity of T (see Remark 3.2.14).

Case 2: $\lambda = q_l$ for some $l \in \{1, \ldots, N\}$. Then $\nabla w|_{\Omega_l} = 0$ and $u|_{\Omega_j} = \frac{q_j}{q_j - \lambda} w|_{\Omega_j} + c_j$ for $j \neq l$ as in case 1. The potential w now has to satisfy the condition (3.21) on the boundary of each inclusion Ω_j except for Ω_l and $\partial_\nu w|_- = 0$ on $\partial \Omega_l$. As in the first case we can represent w by the sum of single layer potentials $w = \sum_{k=1}^{N} SL_k \varphi_k$, and the densities $\varphi_1, \ldots, \varphi_N$ solve

$$\left(-\mu_j I + D_j^* \right) \varphi_j + \sum_{\substack{k=1 \\ k \neq j}}^{N} D_{k \to j}^* \varphi_k = 0 \quad \text{on } \partial \Omega_j \ (j = 1, \ldots, N),$$

where for $j \neq l$ the parameter μ_j is defined as in the first case and $\mu_l = -\frac{1}{2}$.

In both cases a simple computation shows that $\lambda = \frac{2q_j}{2\mu_j q_j + 2 + q_j}$ ($j = 1, \ldots, N$) and thus that the term $\frac{2q_j}{2\mu_j q_j + 2 + q_j}$ is independent of j. Furthermore, we know that $(\varphi_1, \ldots, \varphi_N)^\top \neq (0, \ldots, 0)^\top$ holds since otherwise $\nabla w = \nabla u = 0$.

"$\Longleftarrow$": Let now $\mu_1, \ldots, \mu_N \in \mathbb{R}$ such that there is a nontrivial solution to (3.20) and the term $\frac{2q_j}{2\mu_j q_j + 2 + q_j}$ is identical for all $j = 1, \ldots, N$. As in the first part we distinguish between two different cases. Case 1: $\mu_j \neq -\frac{1}{2}$ for all $j = 1, \ldots, N$. Define $w \in H^1_\diamond(B)$ by $w = \sum_{k=1}^{N} SL_k \varphi_k$, then w solves $\Delta w = 0$ in $B \setminus \partial\Omega$, $\partial_\nu w = 0$ on ∂B and the jump condition (3.21) with $\rho_j = \frac{2\mu_j - 1}{2\mu_j + 1}$ for all $j = 1, \ldots, N$. By assumption we can define $\lambda = \frac{2q_j}{2\mu_j q_j + 2 + q_j}$ for some $j \in \{1, \ldots, N\}$ which is independent of the choice of j, and $u \in H^1(\Omega)$ can be defined by $u|_{\Omega_j} = \frac{q_j}{q_j - \lambda} w|_{\Omega_j}$ since $\lambda \neq q_j$ ($j = 1, \ldots, N$). Now we observe that $\nabla u \in \overline{\mathcal{R}(\mathcal{A})}$, $\mathcal{T}\nabla u = \lambda \nabla u$ and u is non-constant.

Case 2: $\mu_l = -\frac{1}{2}$ for some $l \in \{1, \ldots, N\}$. Now by assumption we can define $\lambda := q_l$. As before define $w = \sum_{k=1}^{N} SL_k \varphi_k$, then $w \in H^1_\diamond(B)$ solves $\Delta w = 0$ in $B \setminus \partial\Omega$, $\partial_\nu w = 0$ on ∂B and the jump condition (3.21) for $\rho_j = \frac{2\mu_j - 1}{2\mu_j + 1}$ and $j \neq l$. Furthermore, $\partial_\nu w|_- = 0$ on $\partial\Omega_l$ which implies $\nabla w|_{\Omega_l} = 0$. We define $u \in H^1(\Omega)$ by $u|_{\Omega_j} = \frac{q_j}{q_j - \lambda} w|_{\Omega_j}$ ($j = 1, \ldots, N, j \neq l$) and $u|_{\Omega_l}$ such that $\Delta u = 0$ in Ω_l and $\partial_\nu u = -\frac{1}{q_l} \partial_\nu w|_+$ on $\partial\Omega_l$. Now we observe that $\nabla u \in \overline{\mathcal{R}(\mathcal{A})}$, u is non-constant on Ω and $\mathcal{T}\nabla u = \lambda \nabla u$, which completes the proof. $\qquad\square$

Remark 3.2.16. In Assumption 3.2.12 it is required that the conductivity contrasts q_j ($j = 1, \ldots, N$) are mutually disjoint from each other. From the correspondence $\lambda = \frac{2q_j}{2\mu_j q_j + 2 + q_j}$ ($j = 1, \ldots, N$) it is easy to see that $\mu_i = \mu_j$ is equivalent to $q_i = q_j$. In order to avoid equality of conductivity contrasts strictly, the claim $\mu_i \neq \mu_j$ for all $i, j = 1, \ldots, N$ such that $i \neq j$ should be included in the formulation of Theorem 3.2.15, while the present formulation of the theorem admits equality of conductivity contrasts. However, in order to keep the assertion simple we omitted this additional condition.

We will also present an assertion according to Theorem 3.2.10 in the case of several inclusions. To this end, the next step is to decouple $\mathcal{T}$ according to the different inclusions, i.e. to find a representation for $\mathcal{T}$ of the type

$$\mathcal{T} = \begin{pmatrix} \mathcal{T}^{(11)} & \cdots & \mathcal{T}^{(1N)} \\ \vdots & \ddots & \vdots \\ \mathcal{T}^{(N1)} & \cdots & \mathcal{T}^{(NN)} \end{pmatrix}$$

such that $\mathcal{T}^{(ij)} : L^2(\Omega_j, \mathbb{R}^d) \to L^2(\Omega_i, \mathbb{R}^d)$. It is easy to check that $\mathcal{T}^{(ii)}$ is defined by $h_i \mapsto q_i (h - \nabla w_i)|_{\Omega_i}$ and that $\mathcal{T}^{(ij)}$ for $i \neq j$ is defined by $h_j \mapsto -q_i \nabla w_j|_{\Omega_i}$. Here, for $h_j \in L^2(\Omega_j, \mathbb{R}^d)$ and $j = 1, \ldots, N$ the potential $w_j \in H^1_\diamond(B)$ is defined as the solution of

$$\iint_B \nabla w_j \cdot \nabla \psi \, dx + \sum_{k=1}^N \iint_{\Omega_k} q_k \nabla w_j \cdot \nabla \psi \, dx = \iint_{\Omega_j} q_j h_j \cdot \nabla \psi \, dx \quad (3.22)$$

for all $\psi \in H^1_\diamond(B)$ (compare the arguments from Section 2.1.2). The following lemma shows a connection to the previously considered case of only one inclusion. We therefore define $\mathcal{T}^{(i)} : L^2(\Omega_i, \mathbb{R}^d) \to L^2(\Omega_i, \mathbb{R}^d)$ as the operator $\mathcal{T}$ that corresponds to the case of only a single inclusion Ω_i, i.e. $\mathcal{T}^{(i)} : h_i \mapsto q_i(h_i - \nabla \tilde{w}_i|_{\Omega_i})$, and $\tilde{w}_i \in H^1_\diamond(B)$ solves

$$\iint_B \nabla \tilde{w}_i \cdot \nabla \psi \, dx + \iint_{\Omega_i} q_i \nabla \tilde{w}_i \cdot \nabla \psi \, dx = \iint_{\Omega_i} q_i h_i \cdot \nabla \psi \, dx \quad (3.23)$$

for all $\psi \in H^1_\diamond(B)$.

Lemma 3.2.17. *Let the operators $\mathcal{T}^{(ij)}$ and $\mathcal{T}^{(i)}$ be defined as above for $i, j = 1, \ldots, N$, then:*

   (a) for $i \neq j$ the operator $\mathcal{T}^{(ij)}$ is compact ($i, j = 1, \ldots, N$),

(b) the operator $S^{(i)} := T^{(ii)} - T^{(i)}$ is compact $(i = 1, \ldots, N)$.

Proof. Part a): The proof of this assertion follows along the same arguments as the proof of Lemma 2.1.8. $T^{(ij)}$ maps $h_j \mapsto -q_i \nabla w_j|_{\Omega_i}$, where w_j is the solution of (3.22). Let $\tilde{\Omega}_i \subset B$ be a simply connected C^2-domain such that $\overline{\Omega}_i \subset \tilde{\Omega}_i$ and $\overline{\tilde{\Omega}}_i \cap \overline{\Omega}_j = \emptyset$ for all $j \neq i$. Then we can decompose $T^{(ij)}$ into $T^{(ij)} = -q_i \tilde{S} \circ \hat{S}$ where $\hat{S} : L^2(\Omega_j, \mathbb{R}^2) \to H_\diamond^{-\frac{1}{2}}(\partial \tilde{\Omega}_i)$ maps h_j to the trace $\partial_\nu w_j|_{\partial \tilde{\Omega}_i}$. $\tilde{S} : H_\diamond^{-\frac{1}{2}}(\partial \tilde{\Omega}_i) \to L^2(\Omega_i, \mathbb{R}^2)$ maps $g \mapsto \nabla v|_{\Omega_i}$ where $v \in H_\diamond^1(\tilde{\Omega}_i)$ solves

$$\Delta v = 0 \text{ in } , \tilde{\Omega}_i \setminus \partial \Omega_i,$$
$$\partial_\nu v = g \text{ on } \partial \tilde{\Omega}_i,$$
$$(1 + q_i) \partial_\nu v|_+ - \partial_\nu v|_- = 0 \text{ on } \partial \Omega_i$$

in the weak sense. Both partial operators are bounded and $\hat{S}$ is compact since $\partial_\nu w|_{\partial \tilde{\Omega}_i} \in H^{\frac{1}{2}}(\partial \tilde{\Omega}_i)$ (compare the proof of Lemma 2.1.8) which proves the first assertion.

Part b): The weak formulation for w_i is (3.22), while the weak formulation for $\tilde{w}_i$ corresponding to $T^{(i)}$, i.e. for the case of the only inclusion Ω_i is (3.23). Setting $v_i := w_i - \tilde{w}_i$ we deduce that v_i solves

$$\iint_B \nabla v_i \cdot \nabla \psi \, dx + \iint_{\Omega_i} q_i \nabla v_i \cdot \nabla \psi \, dx = -\sum_{\substack{k=1 \\ k \neq i}}^{N} \iint_{\Omega_k} q_k \nabla w_i \cdot \nabla \psi \, dx$$

for all $\psi \in H_\diamond^1(B)$. The map $S^{(i)}$ may now be decomposed as follows:

$$h|_{\Omega_i} \mapsto \nabla w_i|_{\Omega \setminus \Omega_i} \mapsto \nabla v_i|_{\Omega_i} \mapsto -q_i \nabla v_i|_{\Omega_i},$$

where we can show analogously to part a) that the first map in this decomposition is compact. Furthermore, it is easy to see that the other maps are bounded.   □

Lemma 3.2.17 now yields the following representation:

$$\mathcal{T} = \begin{pmatrix} T^{(1)} & & 0 \\ & \ddots & \\ 0 & & T^{(N)} \end{pmatrix} + K,$$

with a compact and self-adjoint operator K. This representation is similar
to Corollary 2.1.9 but for the new factorization (3.10) instead of (2.5).

Furthermore, this representation can now be used to investigate the
spectrum of $\mathcal{T}$ as it is done in the following theorem.

Theorem 3.2.18. *In the case of N inclusions $\Omega_1, \ldots, \Omega_N$ with conductivity
contrasts $q_1, \ldots, q_N$ the spectrum of $\mathcal{T}$ consists of a countable set of eigenvalues
and the points $\lambda_1^*, \ldots, \lambda_N^*$ that are defined by $\lambda_j^* = \frac{2q_j}{2+q_j}$ ($j = 1, \ldots, N$). The
points λ_j^* ($j = 1, \ldots, N$) are the only possible accumulation points in $\sigma(\mathcal{T})$.*

Proof. From Theorem 3.2.10 we know that $\mathcal{T}^{(i)} = \lambda_i^* I + K_i$ ($i = 1, \ldots, N$),
and the K_i are self-adjoint and compact operators. Hence we can repre-
sent $\mathcal{T}$ as

$$\mathcal{T} = \begin{pmatrix} \lambda_1^* I & & 0 \\ & \ddots & \\ 0 & & \lambda_N^* I \end{pmatrix} + \tilde{K}, \tag{3.24}$$

and $\tilde{K}$ is a compact operator.

Now we make use of the concept of the essential spectrum (see [48]
or [69]). The essential spectrum consists of all λ for which $\mathcal{T} - \lambda I$ is not
semi-Fredholm, i.e. for these λ neither the nullspace $\mathcal{N}(\mathcal{T} - \lambda I)$ nor the
defect $\overline{\mathcal{R}(\mathcal{A})}/\mathcal{R}(\mathcal{T} - \lambda I)$ is finite-dimensional.

This definition implies immediately that the first part in representation
(3.24), namely the operator

$$\tilde{\mathcal{T}} := \begin{pmatrix} \lambda_1^* I & & 0 \\ & \ddots & \\ 0 & & \lambda_N^* I \end{pmatrix},$$

has the essential spectrum $\{\lambda_1^*, \ldots \lambda_N^*\}$. Theorem 5.35 in Chapter IV of
[48] states that the compact perturbation $\tilde{K}$ has no effect on the essential
spectrum and thus that $\sigma_{ess}(\mathcal{T}) = \{\lambda_1^*, \ldots \lambda_N^*\}$.

From Theorem 5.33 in Chapter IV of [48] we now obtain that $\sigma(\mathcal{T})$
consists of $\sigma_{ess}(\mathcal{T})$ and a countable set of eigenvalues. In addition, these
eigenvalues are isolated eigenvalues, i.e. none of them is an accumula-
tion point in $\sigma(\mathcal{T})$. $\qquad\square$

Remark 3.2.19. It can even be shown that if λ_j^* is an isolated point of $\sigma(\mathcal{T})$, i.e. there is no sequence of eigenvalues converging to λ_j^*, then λ_j^* is itself an eigenvalue of $\mathcal{T}$ with infinite multiplicity (compare Section 3.5 of Chapter V in [48]).

3.2.4 Complex-valued Conductivities

In this section we show that that the previous considerations also apply to complex-valued q. We start by the case of a constant conductivity contrast and assume that $q \in \mathbb{C}$ such that $\operatorname{Re} q \neq 0$ and $\operatorname{Im} q \leq 0$. Note that this assumption on q is also consistent with Assumption 2.1.2.

For complex-valued conductivities the middle operator $\mathcal{T}$ is no longer self adjoint and its spectrum $\sigma(\mathcal{T})$ is part of the complex plane. In fact, $\mathcal{T} : L^2(\Omega, \mathbb{R}^d) \rightarrow L^2(\Omega, \mathbb{R}^d)$ is given by $h \mapsto q(h - \nabla w)$ where $w \in H_\diamond^1(B)$ solves

$$\iint_B (1 + q\chi_\Omega)\nabla w \cdot \nabla \overline{\psi}\, dx = \iint_\Omega qh \cdot \nabla \overline{\psi}\, dx \quad \text{for all } \psi \in H_\diamond^1(B). \quad (3.25)$$

As it is the case for the factorization (2.5) in Chapter 2, the adjoint of the middle operator is given via the complex conjugate of the conductivity.

Lemma 3.2.20. The adjoint $\mathcal{T}^* : L^2(\Omega, \mathbb{R}^d) \rightarrow L^2(\Omega, \mathbb{R}^d)$ of $\mathcal{T}$ is given by $\tilde{h} \mapsto \overline{q}(\tilde{h} - \nabla\tilde{w})$, where $\tilde{w} \in H_\diamond^1(B)$ solves

$$\iint_B (1 + \overline{q}\chi_\Omega)\nabla\tilde{w} \cdot \nabla\overline{\psi}\, dx = \iint_\Omega \overline{q}\tilde{h} \cdot \nabla\overline{\psi}\, dx \quad \text{for all } \psi \in H_\diamond^1(B). \quad (3.26)$$

Proof. Let $h, \tilde{h} \in L^2(\Omega, \mathbb{R}^d)$, and let $w, \tilde{w} \in H_\diamond^1(B)$ be the corresponding solutions of (3.25) and (3.26), respectively. Then:

$$\langle \mathcal{T}h, \tilde{h} \rangle_{L^2(\Omega, \mathbb{R}^d)} = \iint_\Omega q(h - \nabla w) \cdot \overline{\tilde{h}}\, dx$$

$$= \iint_\Omega qh \cdot \overline{\tilde{h}}\, dx - \overline{\iint_B (1 + \overline{q}\chi_\Omega)\nabla\tilde{w} \cdot \nabla\overline{w}\, dx}$$

$$= \iint_\Omega qh \cdot \left(\overline{\tilde{h}} - \nabla\overline{\tilde{w}}\right) dx = \langle h, \mathcal{T}^*\tilde{h} \rangle_{L^2(\Omega, \mathbb{R}^d)}.$$

□

In the following lemma we give bounds for the complex spectrum of T as in Lemma 3.2.1. These bounds show in particular that the operator $\operatorname{Re} T$ is bounded and either positively or negatively coercive.

Lemma 3.2.21. *Let the conductivity contrast $q \in \mathbb{C}$ be such that $|\operatorname{Im} q| < |\operatorname{Re} q|$ and either $\operatorname{Re} q \geq 1$ or $\operatorname{Re} q < 1$ and $|\operatorname{Im} q| < |\operatorname{Re} q| \sqrt{\frac{1+\operatorname{Re} q}{1-\operatorname{Re} q}}$. Then for $\lambda \in \sigma(T)$ it follows that $\operatorname{Re} \lambda \in \left[\frac{\operatorname{Re} q}{1+\operatorname{Re} q}, \operatorname{Re} q \right)$.*

Proof. As in the proof of Lemma 3.2.1 we derive two representations for $\langle Th, h \rangle$ and $\|h\| = 1$:

$$
\begin{aligned}
\langle Th, h \rangle &= \iint_\Omega q\,(h - \nabla w) \cdot \bar{h}\, dx \\
&= \iint_\Omega q\,|h|^2\, dx - \iint_\Omega q\left(\nabla w \cdot \bar{h}\right) dx \qquad (3.27) \\
&= q - \iint_B \frac{q}{\bar{q}}(1 + \bar{q}\chi_\Omega)\,|\nabla w|^2\, dx
\end{aligned}
$$

and

$$
\langle Th, h \rangle = \iint_\Omega q\,|h - \nabla w|^2\, dx + \iint_B |\nabla w|^2\, dx. \qquad (3.28)
$$

The second representation can be derived completely analogously to the proof of Lemma 3.2.1. From (3.27) we deduce

$$
\operatorname{Re} \langle Th, h \rangle = \operatorname{Re} q - \iint_B \operatorname{Re}\left\{ \frac{q}{\bar{q}}(1 + \bar{q}\chi_\Omega) \right\} |\nabla w|^2\, dx,
$$

and it is easy to check that under the above assumptions on q the real part of $\frac{q}{\bar{q}}(1 + \bar{q}\chi_\Omega)$ is positive in the whole of B and thus that $\operatorname{Re} \langle Th, h \rangle <$

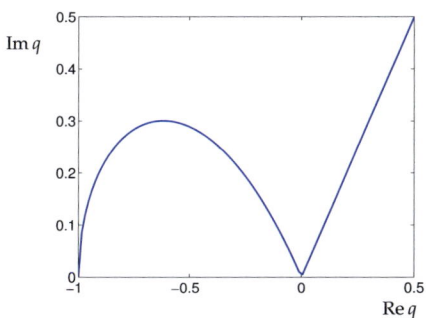

Figure 3.2: Upper bound for $|\mathrm{Im}\,q|$ depending on $\mathrm{Re}\,q$

$\mathrm{Re}\,q$. The remaining bound is derived from (3.28). For $\|h\| = 1$ we obtain

$$\mathrm{Re}\,\langle Th, h\rangle \geq \iint_{\Omega} \left(\mathrm{Re}\,q\,|h|^2 - 2\mathrm{Re}\,q\,\mathrm{Re}\,(h\cdot\nabla\overline{w}) + (1+\mathrm{Re}\,q)\,|\nabla w|^2\right) dx$$

$$= \iint_{\Omega} \left|\sqrt{1+\mathrm{Re}\,q}\,\nabla w - \frac{\mathrm{Re}\,q}{\sqrt{1+\mathrm{Re}\,q}}h\right|^2$$

$$+ \left(\mathrm{Re}\,q - \frac{(\mathrm{Re}\,q)^2}{1+\mathrm{Re}\,q}\right)|h|^2 dx$$

$$\geq \iint_{\Omega} \left(\mathrm{Re}\,q - \frac{(\mathrm{Re}\,q)^2}{1+\mathrm{Re}\,q}\right)|h|^2 dx = \frac{\mathrm{Re}\,q}{1+\mathrm{Re}\,q}.$$

$\square$

Remark 3.2.22. The assumption $|\mathrm{Im}\,q| \leq |\mathrm{Re}\,q|$ and either $\mathrm{Re}\,q \geq 1$ or $\mathrm{Re}\,q < 1$ and $|\mathrm{Im}\,q| \leq |\mathrm{Re}\,q|\sqrt{\frac{1+\mathrm{Re}\,q}{1-\mathrm{Re}\,q}}$ is no severe restriction since in real applications the imaginary part of a complex-valued conductivity is much smaller than the real part. In Figure 3.2 the upper bound for $|\mathrm{Im}\,q|$ depending on $\mathrm{Re}\,q$ for $\mathrm{Re}\,q < 0.5$ is illustrated.

From (3.28) it can easily be seen that $\mathrm{Im}\,\langle Th, h\rangle \leq 0$ since $\mathrm{Im}\,q \leq 0$ and thus that all elements of $\sigma(T)$ have a non-positive imaginary part.

In addition, since T is a bounded operator, there has to be a lower bound for the imaginary part of $\sigma(T)$.

If we look once more at Theorem 3.2.6 and its proof we observe that it doesn't rely on the assumption that q is real-valued. Hence we still have the one-to-one correspondence between the eigenvalues λ of T and the eigenvalues μ of the boundary integral operator D^* via the formula $\mu = -\frac{2+q}{2q} + \frac{1}{\lambda}$. This means in particular that even for complex-valued q existence of eigenvalues of T is assured (see Lemma 3.2.5).

The same holds for Theorem 3.2.10 and complex-valued q since the proof also doesn't rely on real-valued conductivities. This means that $T = \lambda^* I + K$ with a compact operator K and the spectrum of T consists at most of $\lambda^* = \frac{2q}{2+q}$ and a countable set of eigenvalues with the only possible accumulation point λ^*.

Additionally, the theory from Section 3.2.3 for the case N disjoint inclusions $\Omega_1, \ldots, \Omega_N$ can be adapted to complex-valued conductivities. Theorem 3.2.15 and Theorem 3.2.18 may therefore also be applied here.

3.3 Determination of the Conductivity

In this section we explain how the spectrum of T can be obtained from the knowledge of $\Lambda_0 - \Lambda$ and $\mathcal{A}$ and show how it can be used to compute the conductivity contrast q.

We present two different methods to obtain an approximation of $\sigma(T)$ out of which the first one provides a one-to-one connection between $\sigma(T)$ and the spectrum of an operator that can be constructed from measured data. The proof of this connection is closely related to the proof that the Factorization method works and therefore it relies on the coercivity of T.

The second method is based on projection on finite-dimensional subspaces and provides only an asymptotic result for eigenvalues of the finite-dimensional problem. However, it doesn't rely on coercivity of T and can therefore even be applied to the mixed problem from Chapter 2.

3.3.1 Approximation of the Spectrum

Our first method to obtain the spectrum $\sigma(\mathcal{T})$ is based on the factorization (3.10), and it relies on the following assumption.

Assumption 3.3.1. Assume that that q is is constant and real-valued or, for N inclusions, either $q_j > 0$ for all $j = 1, \ldots, N$ or $q_j < 0$ for all $j = 1, \ldots, N$.

In this case we know that either $\Lambda_0 - \Lambda$ is either positive definite or negative definite which implies that the operator $|\Lambda_0 - \Lambda| : L^2_\diamond(\partial B) \to L^2_\diamond(\partial B)$ is positive definite and self-adjoint. This implies the existence of the square root operator $|\Lambda_0 - \Lambda|^{\frac{1}{2}} : L^2_\diamond(\partial B) \to L^2_\diamond(\partial B)$. In addition, the main result of the Factorization method is the range identity $\mathcal{R}(|\Lambda_0 - \Lambda|^{\frac{1}{2}}) = \mathcal{R}(\mathcal{A}^*)$ (see [55]). We denote this range by Y and note that it is a dense subspace of $L^2_\diamond(\partial B)$.

As a motivation of our procedure consider the generalized eigenvalue problem

$$|\Lambda_0 - \Lambda| f = \lambda \mathcal{A}^* \mathcal{A} f \qquad (3.29)$$

for $f \in L^2_\diamond(\partial B)$. Using the factorization (3.10), (3.29) can be rewritten as

$$\mathcal{A}^* \left(|\mathcal{T}| - \lambda I \right) \mathcal{A} f = 0.$$

Now we observe that there is an eigenpair $\{\lambda, f\}$ of (3.29) if and only if λ is an eigenvalue of $|\mathcal{T}|$ and at least one corresponding eigenfunction h is contained in $\mathcal{R}(\mathcal{A})$. The existence of solutions to the generalized eigenvalue problem is not clear since we only investigated $\sigma(\mathcal{T})$ on the closed space $\overline{\mathcal{R}(\mathcal{A})}$, and even if there are such generalized eigenvalues λ we cannot hope to find all eigenvalues of $\mathcal{T}$ using (3.29).

In the following considerations we derive a new representation of the generalized eigenvalue problem that yields a one-to-one connection to $\sigma(\mathcal{T})$. Using the substitution $g := |\Lambda_0 - \Lambda|^{\frac{1}{2}} f$ and applying $|\Lambda_0 - \Lambda|^{-\frac{1}{2}}$ to (3.29), we obtain

$$\frac{1}{\lambda} g = |\Lambda_0 - \Lambda|^{-\frac{1}{2}} \mathcal{A}^* \mathcal{A} |\Lambda_0 - \Lambda|^{-\frac{1}{2}} g,$$

which is an ordinary eigenvalue problem. Nevertheless, g has to be contained in Y and existence of eigenvalues is still not assured. In the following we thus extend this new eigenvalue problem to $L^2_\diamond(\partial B)$.

Let us define the operator $\mathcal{B} : Y \to \overline{\mathcal{R}(\mathcal{A})}$ by

$$\mathcal{B} = \mathcal{A} \, |\Lambda_0 - \Lambda|^{-\frac{1}{2}}$$

as well as the operator $\tilde{\mathcal{B}} : \overline{\mathcal{R}(\mathcal{A})} \to L^2_\diamond(\partial B)$ by

$$\tilde{\mathcal{B}} = |\Lambda_0 - \Lambda|^{-\frac{1}{2}} \, \mathcal{A}^*.$$

Using these operators we derive a connection to the spectrum of $\mathcal{T}$. However, we first have to show some properties of $\mathcal{B}$ and $\tilde{\mathcal{B}}$. By construction of $\mathcal{B}$ and $\tilde{\mathcal{B}}$ it is easy to see that

$$\tilde{\mathcal{B}} \, |\mathcal{T}| \, \mathcal{B} = |\Lambda_0 - \Lambda|^{-\frac{1}{2}} \, \mathcal{A}^* \, |\mathcal{T}| \, \mathcal{A} \, |\Lambda_0 - \Lambda|^{-\frac{1}{2}} = I_Y \qquad (3.30)$$

holds. In the following lemma we show that $\mathcal{B}, \tilde{\mathcal{B}}$ are boundedly invertible and adjoint to each other.

Lemma 3.3.2. *Let Assumption 3.3.1 hold. Then the operators $\mathcal{B}$, $\tilde{\mathcal{B}}$ have the following properties:*

(a) *$\mathcal{B}$ has a bounded extension $\mathcal{B} : L^2_\diamond(\partial B) \to \overline{\mathcal{R}(\mathcal{A})}$.*

(b) *$\tilde{\mathcal{B}} : \overline{\mathcal{R}(\mathcal{A})} \to L^2_\diamond(\partial B)$ is also bounded.*

(c) *$\mathcal{B}$ and $\tilde{\mathcal{B}}$ are adjoint to each other.*

(d) *$\mathcal{B} : L^2_\diamond(\partial B) \to \overline{\mathcal{R}(\mathcal{A})}$ and $\tilde{\mathcal{B}} : \overline{\mathcal{R}(\mathcal{A})} \to L^2_\diamond(\partial B)$ are boundedly invertible and $\tilde{\mathcal{B}} \, |\mathcal{T}| \, \mathcal{B} = I_{L^2_\diamond(\partial B)}$.*

Proof. For arbitrary $x \in Y$ and $h \in \overline{\mathcal{R}(\mathcal{A})}$ we have that

$$
\begin{aligned}
\langle \mathcal{B}x, h \rangle_{L^2(\Omega, \mathbb{R}^d)} &= \left\langle \mathcal{A} \, |\Lambda_0 - \Lambda|^{-\frac{1}{2}} x, h \right\rangle_{L^2(\Omega, \mathbb{R}^d)} \\
&= \left\langle |\Lambda_0 - \Lambda|^{-\frac{1}{2}} x, \mathcal{A}^* h \right\rangle_{L^2(\partial B)} \qquad (3.31) \\
&= \left\langle x, |\Lambda_0 - \Lambda|^{-\frac{1}{2}} \mathcal{A}^* h \right\rangle_{L^2(\partial B)} = \langle x, \tilde{\mathcal{B}} h \rangle_{L^2(\partial B)}.
\end{aligned}
$$

However, it is not yet shown that $\tilde{\mathcal{B}}, \mathcal{B}$ are adjoint to each other since $\mathcal{B}x$ is only defined for $x \in Y$.

The operator $|\mathcal{T}| : \overline{\mathcal{R}(\mathcal{A})} \rightarrow \overline{\mathcal{R}(\mathcal{A})}$ is bounded, self-adjoint and coercive, i.e. there are constants $c_1, c_2 > 0$ such that

$$c_1 \|h\|^2_{L^2(\Omega,\mathbb{R}^d)} \leq \langle |\mathcal{T}| h, h \rangle \leq c_2 \|h\|^2_{L^2(\Omega,\mathbb{R}^d)} \quad \text{for all } h \in \overline{\mathcal{R}(\mathcal{A})}.$$

Now let $x \in Y$ such that $\|x\| = 1$ then we know from (3.30) that

$$1 = \langle \tilde{\mathcal{B}} |\mathcal{T}| \mathcal{B}x, x \rangle_{L^2(\partial B)} = \langle |\mathcal{T}| \mathcal{B}x, \mathcal{B}x \rangle_{L^2(\Omega,\mathbb{R}^d)} \geq c_1 \|\mathcal{B}x\|^2_{L^2(\partial B)}.$$

We have thus proven that $\mathcal{B}$ is bounded on Y, which also implies that $\mathcal{B}$ has a bounded extension $\mathcal{B} : L^2_\diamond(\partial B) \rightarrow \overline{\mathcal{R}(\mathcal{A})}$. Furthermore, (3.31) holds for all $x \in L^2_\diamond(\partial B)$ and thus $\mathcal{B}, \tilde{\mathcal{B}}$ are adjoint to each other. This implies that $\tilde{\mathcal{B}}$ is also bounded on $\overline{\mathcal{R}(\mathcal{A})}$ and (3.30) can be extended to $\tilde{\mathcal{B}} |\mathcal{T}| \mathcal{B} = I_{L^2_\diamond(\partial B)}$.

This equality can now be rewritten as $\tilde{\mathcal{B}} |\mathcal{T}|^{\frac{1}{2}} |\mathcal{T}|^{\frac{1}{2}} \mathcal{B} = I_{L^2_\diamond(\partial B)}$ which implies $\left(|\mathcal{T}|^{\frac{1}{2}} \mathcal{B} \right)^* \left(|\mathcal{T}|^{\frac{1}{2}} \mathcal{B} \right) = I_{L^2_\diamond(\partial B)}$. This implies that $\left(|\mathcal{T}|^{\frac{1}{2}} \mathcal{B} \right)$ is one-to-one and that $\left(|\mathcal{T}|^{\frac{1}{2}} \mathcal{B} \right)^*$ is onto. By construction $\mathcal{B}$ has dense range in $\overline{\mathcal{R}(\mathcal{A})}$, and $|\mathcal{T}|^{\frac{1}{2}} : \overline{\mathcal{R}(\mathcal{A})} \rightarrow \overline{\mathcal{R}(\mathcal{A})}$ is an isomorphism. Hence $\left(|\mathcal{T}|^{\frac{1}{2}} \mathcal{B} \right)$ has dense range and $\left(|\mathcal{T}|^{\frac{1}{2}} \mathcal{B} \right)^*$ is also one-to-one. This proves that $\left(|\mathcal{T}|^{\frac{1}{2}} \mathcal{B} \right)^*$ is boundedly invertible, and the same holds for $|\mathcal{T}|^{\frac{1}{2}} \mathcal{B}$. Since $|\mathcal{T}|^{\frac{1}{2}}$ is boundedly invertible, $\mathcal{B}$ is also boundedly invertible and the same holds for $\tilde{\mathcal{B}}$. $\square$

Now we return to the investigation of the spectrum of $\mathcal{T}$ and show a connection to the operators $\mathcal{B}$ and $\tilde{\mathcal{B}}$.

Corollary 3.3.3. *Let $\lambda \in \mathbb{R}$, then $\lambda \in \sigma(|\mathcal{T}|)$ if and only if $\frac{1}{\lambda} \in \sigma(\tilde{\mathcal{B}}\mathcal{B})$.*

Proof. Since both $|\mathcal{T}|$ and $\tilde{\mathcal{B}}\mathcal{B}$ are boundedly invertible operators, their spectra don't contain 0. Let $0 \neq \mu$ be in the resolvent set $\varrho(\tilde{\mathcal{B}}\mathcal{B})$, i.e. the operator $R := \mu I - \tilde{\mathcal{B}}\mathcal{B}$ is boundedly invertible. From Lemma 3.3.2 we know that

$$R = \mu \tilde{\mathcal{B}} |\mathcal{T}| \mathcal{B} - \tilde{\mathcal{B}}\mathcal{B} = \tilde{\mathcal{B}}(\mu |\mathcal{T}| - I)\mathcal{B}.$$

Hence R is boundedly invertible if and only if $\lambda := \frac{1}{\mu} \in \varrho(|\mathcal{T}|)$, since $\mathcal{B}, \tilde{\mathcal{B}}$ are boundedly invertible as well. This also implies that the spectra $\sigma(\tilde{\mathcal{B}}\mathcal{B})$ and $\sigma(|\mathcal{T}|)$ are connected via the relation $\mu = \frac{1}{\lambda}$. $\qquad\qquad\square$

The operator $\tilde{\mathcal{B}}\mathcal{B}$ can principally be obtained from measured data and a priori information and its spectrum can therefore be used to compute the spectrum of $\mathcal{T}$.

The second possibility to compute an approximation to $\sigma(\mathcal{T})$ is to make use of the generalized eigenvalue problem (3.29). We show that for a discrete version of (3.29) existence of solutions is assured and that a subsequence of those generalized eigenvalues converges towards an eigenvalue of $\mathcal{T}$ or an accumulation point in $\sigma(\mathcal{T})$. This method can be used independently of whether Assumption 3.3.1 holds or not.

Let Y_n be an n-dimensional subspace of $L^2_\diamond(\partial B)$ and let $X_n = \mathcal{A}(Y_n)$. X_n is also n-dimensional since $\mathcal{A}$ is injective. We define the maps $P_n : L^2_\diamond(\partial B) \to Y_n$, $Q_n : L^2(\Omega, \mathbb{R}^d) \to X_n$ as the corresponding orthogonal projections. From the factorization (3.10) we obtain its finite-dimensional counterpart by orthogonal projection:

$$P_n(\Lambda_0 - \Lambda)|_{Y_n} = P_n \mathcal{A}^* \mathcal{T} \mathcal{A}|_{Y_n}.$$

Now we regard the generalized eigenvalue problem

$$P_n(\Lambda_0 - \Lambda)f_n = \lambda_n P_n \mathcal{A}^* \mathcal{A} f_n \quad (f_n \in Y_n), \qquad (3.32)$$

which has a set of n possibly complex-valued eigenvalues $\lambda_n^{(1)}, \ldots, \lambda_n^{(n)}$ and corresponding eigenfunctions $f_n^{(1)}, \ldots, f_n^{(n)}$, since injectivity of $\mathcal{A}$ implies injectivity of $P_n \mathcal{A}^* \mathcal{A}|_{Y_n}$. If γ is real-valued then $\Lambda_0 - \Lambda$ is self-adjoint which also implies self-adjointness of $P_n(\Lambda_0 - \Lambda)|_{Y_n}$. The operator $P_n \mathcal{A}^* \mathcal{A}|_{Y_n}$ is self-adjoint and positive definite independently of the conductivity. Hence the eigenvalues according to (3.32) are real-valued if γ is real-valued.

We consider the behavior of $\lambda_n^{(j)}$ and $f_n^{(j)}$ for $n \to \infty$ and arbitrary $j \in \mathbb{N}$. In order to keep our notations simple we only write λ_n, f_n instead of $\lambda_n^{(j)}$ and $f_n^{(j)}$, where λ_n denotes $\lambda_n^{(j)}$ for some $j \in \mathbb{N}$.

In the following lemmas we show that the discrete eigenvalue problem (3.32) can be used to approximate the spectrum of $\mathcal{T}$. We distinguish between the case of a constant conductivity contrast as in Section 3.2.2 and the case of N different contrasts $q_1, \ldots, q_N$ as in Section 3.2.3. In particular, in the second case we don't need Assumption 3.3.1 to hold which makes this method especially attractive in the mixed case from Chapter 2. However, we first consider the case of a single conductivity contrast q and we denote $\lambda^* = \frac{2q}{2+q}$ as in Section 3.2.2.

Lemma 3.3.4. *Let q be constant and let λ_n be eigenvalues according to the discrete generalized eigenvalue problem (3.32) for $n \in \mathbb{N}$. Then there exists a convergent subsequence $(\lambda_{n_k})_{k\in\mathbb{N}}$ with limit λ that is an eigenvalue of $\mathcal{T}$ or equal to λ^*.*

In addition, if the sequence $(\lambda_n)_{n\in\mathbb{N}}$ itself converges to some λ, then the limit λ is an eigenvalue of $\mathcal{T}$ or equal to λ^*.

Proof. We first derive an equivalent formulation to the eigenvalue problem (3.32) that can be used to show convergence of eigenvalues towards eigenvalues of $\mathcal{T}$ and the connection to the infinite-dimensional case. By the factorization of (3.10) and Theorem 3.2.10, (3.32) can be written as

$$\lambda^* P_n \mathcal{A}^* \mathcal{A} f_n + P_n \mathcal{A}^* K \mathcal{A} f_n = \lambda_n P_n \mathcal{A}^* \mathcal{A} f_n.$$

Setting $g_n := \mathcal{A} f_n$ and assuming without loss of generality that $\|g_n\| = 1$ we write

$$(\lambda^* - \lambda_n) P_n \mathcal{A}^* g_n + P_n \mathcal{A}^* K g_n = 0.$$

This means that we have

$$\left\langle (\lambda^* - \lambda_n) \mathcal{A}^* g_n, \psi_n \right\rangle_{L^2(\partial B)} + \left\langle \mathcal{A}^* K g_n, \psi_n \right\rangle_{L^2(\partial B)} = 0$$

for all $\psi_n \in Y_n$. Using the duality of $\mathcal{A}$ and $\mathcal{A}^*$ we obtain

$$\left\langle (\lambda^* - \lambda_n) g_n, \mathcal{A}\psi_n \right\rangle_{L^2(\Omega, \mathbb{R}^d)} + \left\langle K g_n, \mathcal{A}\psi_n \right\rangle_{L^2(\Omega, \mathbb{R}^d)} = 0$$

for all $\psi_n \in Y_n$ and thus

$$\left\langle (\lambda^* - \lambda_n) g_n, \phi_n \right\rangle_{L^2(\Omega, \mathbb{R}^d)} + \left\langle K g_n, \phi_n \right\rangle_{L^2(\Omega, \mathbb{R}^d)} = 0$$

for all $\phi_n \in X_n$. Using the orthogonal projection Q_n we obtain

$$(\lambda^* - \lambda_n) g_n + Q_n K g_n = 0. \tag{3.33}$$

Let λ_n, g_n be eigenpairs according to (3.33) for $n \in \mathbb{N}$. Since the g_n are uniformly bounded there is a subsequence g_{n_k} that converges weakly towards some $g \in L^2(\Omega, \mathbb{R}^d)$ for $k \to \infty$. We know by compactness of K that $Q_{n_k} K g_{n_k} \to Kg$ ($k \to \infty$). This means that $(\lambda^* - \lambda_{n_k}) g_{n_k} \to -Kg$ ($k \to \infty$) and thus

$$\left\langle \left(\lambda^* - \lambda_{n_k}\right) g_{n_k}, \phi \right\rangle_{L^2(\Omega, \mathbb{R}^d)} \to - \left\langle Kg, \phi \right\rangle_{L^2(\Omega, \mathbb{R}^d)} \quad (k \to \infty).$$

for all $\phi \in L^2(\Omega, \mathbb{R}^d)$.

We first assume that $g \neq 0$, then the weak convergence yields that $\langle g_{n_k}, \phi \rangle_{L^2(\Omega, \mathbb{R}^d)} \to \langle g, \phi \rangle_{L^2(\Omega, \mathbb{R}^d)}$ for all $\phi \in L^2(\Omega, \mathbb{R}^d)$ and thus

$$\lambda_{n_k} \left\langle g_{n_k}, \phi \right\rangle_{L^2(\Omega, \mathbb{R}^d)} \to \lambda^* \left\langle g, \phi \right\rangle_{L^2(\Omega, \mathbb{R}^d)} + \left\langle Kg, \phi \right\rangle_{L^2(\Omega, \mathbb{R}^d)}$$
$$= \left\langle Tg, \phi \right\rangle_{L^2(\Omega, \mathbb{R}^d)}.$$

By setting $\phi = g$ and using $g \neq 0$ we obtain $\lambda_{n_k} \to \lambda$ ($k \to \infty$) for some $\lambda \in \mathbb{R}$, which implies $Tg = \lambda g$, i.e. λ is an eigenvalue of T.

Now assume that $g = 0$, hence $Kg = 0$ and $(\lambda^* - \lambda_{n_k}) g_{n_k} \to 0$ ($k \to \infty$). If $\lambda_{n_k} \not\to \lambda^*$, then for a subsequence $g_{n_{k_l}}$ we have $g_{n_{k_l}} \to 0$, in contradiction to $\|g_n\| = 1$. Hence $\lambda_{n_k} \to \lambda^*$ ($k \to \infty$).

For the second assertion, assume that $\lambda_n \to \lambda$ ($n \to \infty$) and $\lambda \neq \lambda^*$. As before, there is a weakly convergent subsequence $g_{n_k} \rightharpoonup g$ of the corresponding generalized eigenfunctions according to (3.33). From (3.33) we deduce that $Q_{n_k} K g_{n_k} \to Kg$ and hence $Tg = \lambda g$ and $g \neq 0$ as shown above, which completes the proof.

$\square$

This result means that for sufficiently large n at least some of the generalized eigenvalues λ_n of the discrete eigenvalue problem (3.32) lie arbitrarily close to eigenvalues of T or to λ^*. In particular, we observe that they accumulate in λ^*.

An optimal result in this context would be that for n sufficiently large all the eigenvalues from (3.29) are lying in a neighborhood of elements of $\sigma(T)$ and that every element of $\sigma(T)$ is approximated by eigenvalues from (3.29). However, this question will be subject to future work.

We proceed by applying the projection method to the case of N inclusions. As before, let Y_n be an n-dimensional subspace of $L^2_\diamond(\partial B)$ and

$X_n = \mathcal{A}(Y_n)$. The orthogonal projections P_n, Q_n are defined as in the case of only one inclusion, and we denote $\lambda_i^* = \frac{2q_i}{2+q_i}$ for $(i = 1, \ldots, N)$ as is Section 3.2.3.

Lemma 3.3.5. *Let $q_1, \ldots, q_N$ be as in Section 3.2.3 and let λ_n be eigenvalues according to the discrete generalized eigenvalue problem (3.32) for $n \in \mathbb{N}$. Then there exists a convergent subsequence $(\lambda_{n_k})_{k \in \mathbb{N}}$ with limit λ that is an eigenvalue of $\mathcal{T}$ or equal to λ_j^* for some $j \in \{1, \ldots, N\}$.*

In addition, if the sequence $(\lambda_n)_{n \in \mathbb{N}}$ itself converges to some λ, then the limit λ is an eigenvalue of $\mathcal{T}$ or equal to λ_j^* for some $j \in \{1, \ldots, N\}$.

Proof. The proof is very similar to the proof of Lemma 3.3.4. Our starting point is again the discrete generalized eigenvalue problem (3.32) which is equivalent to

$$P_n \mathcal{A}^* \begin{pmatrix} (\lambda_1^* - \lambda_n) I & & 0 \\ & \ddots & \\ 0 & & (\lambda_N^* - \lambda_n) I \end{pmatrix} \mathcal{A} f_n + P_n \mathcal{A}^* K \mathcal{A} f_n = 0,$$

where we used (3.10) and the proof of Theorem 3.2.18. This problem can be transformed the same way as we did for Lemma 3.3.4 to the problem

$$\begin{pmatrix} (\lambda_1^* - \lambda_n) I & & 0 \\ & \ddots & \\ 0 & & (\lambda_N^* - \lambda_n) I \end{pmatrix} g_n + Q_n K g_n = 0, \tag{3.34}$$

where we set $g_n = \mathcal{A} f_n$ and $\|g_n\| = 1$ without loss of generality.

Let λ_n be eigenvalues according to (3.34) for $n \in \mathbb{N}$ and let $g_{n_k} \rightharpoonup g$ be a weakly convergent subsequence of the corresponding generalized eigenfunctions g_n. Then $Q_{n_k} K g_{n_k} \to K g$ and

$$\begin{pmatrix} (\lambda_1^* - \lambda_{n_k}) I & & 0 \\ & \ddots & \\ 0 & & (\lambda_N^* - \lambda_{n_k}) I \end{pmatrix} g_{n_k} \to -K g \quad (k \to \infty).$$

In the present case g_{n_k} consists of N components $g_{n_k}^{(1)}, \ldots, g_{n_k}^{(N)}$ corresponding to the inclusions $\Omega_1, \ldots, \Omega_N$. As before we have to consider the two different cases $g \neq 0$ and $g = 0$. First assume that $g \neq 0$. We know that

$$\sum_{i=1}^{N} \left\langle \left(\lambda_i^* - \lambda_{n_k} \right) g_{n_k}^{(i)}, \phi^{(i)} \right\rangle_{L^2(\Omega_i, \mathbb{R}^d)} \rightarrow - \langle Kg, \phi \rangle_{L^2(\partial B)} \quad (k \rightarrow \infty)$$

with $\phi^{(i)} = \phi|_{\Omega_i}$ $(i = 1, \ldots, N)$ holds for all $\phi \in L^2(\Omega, \mathbb{R}^d)$. From the convergence $\langle g_{n_k}, \phi \rangle_{L^2(\Omega, \mathbb{R}^d)} \rightarrow \langle g, \phi \rangle_{L^2(\Omega, \mathbb{R}^d)}$ for all $\phi \in L^2(\Omega, \mathbb{R}^d)$ we obtain

$$\lambda_{n_k} \langle g_{n_k}, \phi \rangle_{L^2(\Omega, \mathbb{R}^d)} \rightarrow \sum_{i=1}^{N} \left\langle \lambda_i^* g^{(i)}, \phi^{(i)} \right\rangle_{L^2(\Omega, \mathbb{R}^d)} + \langle Kg, \phi \rangle_{L^2(\Omega, \mathbb{R}^d)}$$

$$= \langle \mathcal{T}g, \phi \rangle_{L^2(\Omega, \mathbb{R}^d)}.$$

By setting $\phi = g$ and using $g \neq 0$ we obtain $\lambda_{n_k} \rightarrow \lambda$ $(k \rightarrow \infty)$ for some $\lambda \in \mathbb{R}$ which implies $\mathcal{T}g = \lambda g$ and λ is an eigenvalue of $\mathcal{T}$.

Now assume that $g = 0$, hence $Kg = 0$ and

$$\left(\lambda_i^* - \lambda_{n_k} \right) g_{n_k}^{(i)} \rightarrow 0 \quad (i = 1, \ldots, N).$$

If $\lambda_{n_k} \nrightarrow \lambda_i^*$ for all $i \in \{1, \ldots, N\}$ then there is a subsequence $g_{n_{k_l}} \rightarrow 0$ $(l \rightarrow \infty)$, a contradiction to $\|g_n\| = 1$. Hence $\lambda_{n_k} \rightarrow \lambda_i^*$ for some $i \in \{1, \ldots, N\}$.

In addition, we can show that if a sequence $(\lambda_n)_{n \in \mathbb{N}}$ of the generalized eigenvalues converges to some λ, then λ is an eigenvalue of $\mathcal{T}$ or equal to one of the accumulation points λ_i^*. The corresponding proof is also analogous to the case of one inclusion.

$\square$

These results show that the discrete generalized eigenvalue problem (3.32) may be used to approximate $\sigma(\mathcal{T})$. However, we have not shown that this method provides all eigenvalues of $\mathcal{T}$ as n tends to infinity. One advantage of this method is that it also works for mixed inclusions, i.e. for the case in which we have both inclusions with a lower and inclusions with a higher conductivity than the background. In this case $\mathcal{T}$ is no longer positively or negative coercive, and the proof of Lemma 3.3.2 does not work.

3.3.2 Approximation of the Conductivity

Once the spectrum of $\mathcal{T}$ is obtained, the next step is to compute the conductivity contrast q (the contrasts $q_1, \ldots, q_N$ in the case of N inclusions, respectively). There are several methods for this problem, and we start by explaining the most obvious ones.

First of all, we know from Lemma 3.2.1 that an upper bound and and a lower bound for the spectrum of $\mathcal{T}$. From these we can obtain bounds for the conductivity contrast q (for $q_1, \ldots, q_N$, respectively) using the largest and the smallest eigenvalue of $\mathcal{T}$. Analogously, we can use Lemma 3.2.13 to obtain bounds for N different conductivity contrasts $q_1, \ldots, q_N$. If γ is also complex-valued, we can make use of Lemma 3.2.21. However, this method doesn't give exact values for the conductivities and we therefore turn to more exact methods.

In Theorem 3.2.10 we have proven that in the case of a constant conductivity contrast q the eigenvalues of $\mathcal{T}$ exhibit exactly one accumulation point, namely $\lambda^* = \frac{2q}{2+q}$. We thus have to identify this limit point in the numerically approximated spectrum of $\mathcal{T}$ and to compute q from this value. In the case of N disjoint inclusions $\sigma(\mathcal{T})$ has N possible accumulation points $\lambda_1^*, \ldots, \lambda_N^*$ (compare Theorem 3.2.18), and we can compute the conductivity contrasts $q_1, \ldots, q_N$ using the relation $\lambda_j^* = \frac{2q_j}{2+q_j}$ ($j = 1, \ldots, N$).

For constant q this method is recommendable since it is fast and accurate. However, for several inclusions with different conductivity contrasts $q_1, \ldots, q_N$ it remains open which conductivity contrast belongs to which inclusion. In some applications this assignment may be available from a priori information, but otherwise we have so solve the direct problem in order to find the correct conductivity assignment.

In Corollary 3.2.11 we showed a direct connection between the spectrum of $\mathcal{T}$ and the spectrum of the boundary integral operator D^*. This connection provides another method to compute q since D^* and its eigenvalues μ_j ($j \in \mathbb{N}$) can be computed from the knowledge about the inclusion boundary $\partial\Omega$ and hence without the knowledge of q. Having found the eigenvalues λ_j ($j \in \mathbb{N}$) of $\mathcal{T}$, we can compute q from the correspondence $\mu_j = -\frac{2+q}{2q} + \frac{1}{\lambda_j}$.

However, we cannot use this procedure in the case of N inclusions.

The reason is that according to Theorem 3.2.15 we have to find $\mu_1, \ldots, \mu_N$ such that (3.20) has a nontrivial solution under the additional condition that the value of the term $\frac{2q_j}{2\mu_j q_j + 2 + q_j}$ doesn't depend on j for $j = 1, \ldots, N$. Without this condition there is in general an uncountable set of tuples $\mu_1, \ldots, \mu_N$ such that the integral equation system (3.20) has nontrivial solutions. But the condition that $\frac{2q_j}{2\mu_j q_j + 2 + q_j} = const$ is obviously depending on the unknowns $q_1, \ldots, q_N$.

3.4 Numerical Experiments

In this section we show some numerical examples with our new method to approximate the conductivity inside anomalies. We first explain how the operators $\Lambda_0 - \Lambda$ and $\mathcal{A}^*\mathcal{A}$ are computed numerically and discuss several possibilities to obtain approximations of $\sigma(\mathcal{T})$ and q. Afterwards we present some examples concerning exact and inexact data $\Lambda_0 - \Lambda$ as well as exact and inexact inclusion boundaries.

3.4.1 Numerical Solution of the Direct Problem

As in Section 2.3, in all our examples the domain B is the unit disc in $\mathbb{R}^2$. At first we need to compute discrete versions of $\Lambda_0 - \Lambda$ and of $\mathcal{A}^*\mathcal{A}$. Since we restrict ourselves to piecewise constant conductivities these operators can be computed easily using boundary integral equation methods.

Let us fix some arbitrary current pattern $f \in L^2_\diamond(\partial B)$. Now define $v \in H^1_\diamond(B)$ by $v := u_0 - u$ where $u \in H^1_\diamond(B)$ solves the direct problem (1.5) with the conductivity $\gamma(x) = 1 + q\chi_\Omega(x)$ and $u_0 \in H^1_\diamond(B)$ solves (3.2). Then for v we obtain the transmission problem

$$\Delta v = 0 \text{ in } B \setminus \partial\Omega,$$
$$\partial_\nu v = 0 \text{ on } \partial B, \tag{3.35}$$
$$(1 + q)\,\partial_\nu v|_- - \partial_\nu v|_+ = q\,\partial_\nu u_0 \text{ on } \partial\Omega.$$

For B being the unit disc, u_0 may be calculated explicitly, and (3.35) can be transformed to a boundary integral equation of second type. By setting

$v = SL\varphi$ we obtain the following integral equation on $\partial\Omega$ for the density $\varphi \in H_\diamond^{-\frac{1}{2}}(\partial\Omega)$:

$$\left(1 + \frac{q}{2}\right)\varphi + qD^*\varphi = q\,\partial_\nu u_0.$$

For $\mathcal{A}^*\mathcal{A}$ the corresponding transmission problem is (3.7) with $h = \nabla u_0$. Here it is not even necessary to solve a boundary integral equation as for $\Lambda_0 - \Lambda$ but it is sufficient to define $v = SL\psi$ with $\psi = \partial_\nu u_0|_{\partial\Omega}$.

Using these integral equation methods for n orthogonal current patterns as in (2.33) we obtain the discrete versions $L_n := P_n(\Lambda_0 - \Lambda)|_{Y_n}$ and $A_n := P_n\mathcal{A}^*\mathcal{A}|_{Y_n}$ of $\Lambda_0 - \Lambda$ and $\mathcal{A}^*\mathcal{A}$, where Y_n is the n-dimensional subspace of $L_\diamond^2(\partial B)$ spanned by the basis of current patterns.

3.4.2 Approximation of Spectrum and Conductivity

We use two different test models, where in the first test model there is an elliptic inclusion Ω in the upper right part of B having the conductivity contrast $q = 1$. In the second test model there are two inclusions: Ω_1 is the same inclusion as Ω in the first model, i.e. also with $q_1 = 1$, while Ω_2 is a circle located in the lower left part of B with $q_2 = -0.5$. This means in particular that test model 2 corresponds to the case of mixed inclusions from Chapter 2. The inclusions are illustrated in Figure 3.3.

As described in Section 3.3.1, there are two possibilities to calculate $\sigma(\mathcal{T})$ numerically. The first one is to compute a discrete version B_n of the operator $\check{\mathcal{B}}\mathcal{B}$, to conduct an eigenvalue decomposition of B_n and to invert these eigenvalues (compare Corollary 3.3.3). For the discrete version B_n of $\check{\mathcal{B}}\mathcal{B}$ we use the composition $B_n = L_n^{-\frac{1}{2}} A_n L_n^{-\frac{1}{2}}$.

The second possibility is to make use of the generalized eigenvalue problem (3.32) that can be rewritten as

$$L_n f_n = \lambda_n A_n f_n \quad (f_n \in Y_n). \tag{3.36}$$

Such a generalized eigenvalue problem can be solved using the Cholesky factorization (see e.g. [74]). Since the conductivity is real-valued in both test models the operator $\Lambda_0 - \Lambda$ is self-adjoint. This implies that the operators L_n and A_n are also self-adjoint und hence that the eigenvalues according to (3.36) are real-valued.

Our first numerical example consists of a comparison between these two methods. Figure 3.4 shows the approximated spectrum of $\mathcal{T}$ for both

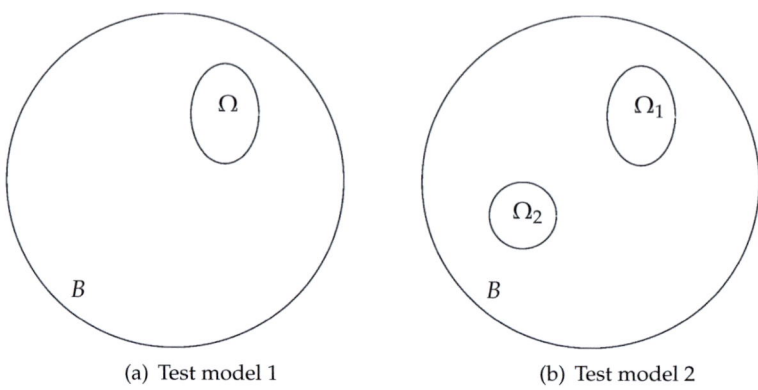

(a) Test model 1   (b) Test model 2

Figure 3.3: The two test models with corresponding inclusions

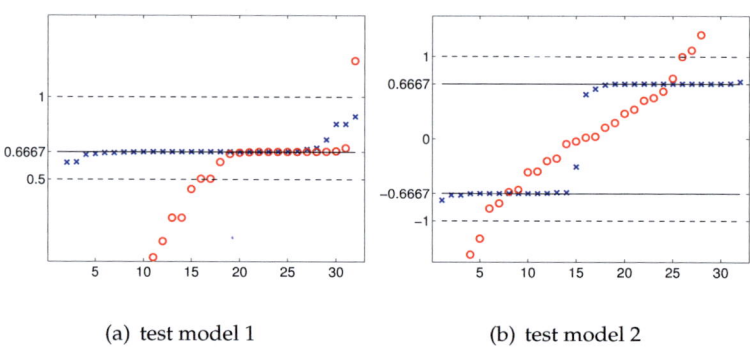

(a) test model 1   (b) test model 2

Figure 3.4: Approximation of $\sigma(\mathcal{T})$ for both test models. blue '$\times$': generalized eigenvalue problem (3.36), red '$\circ$': use of B_n, continuous line: the limit point λ^* (the accumulation points λ_1^*, λ_2^*, respectively), dashed lines: upper and lower bound for $\sigma(\mathcal{T})$

test models. The blue '×' indicate the eigenvalues obtained from (3.36) while the red 'o' correspond to the inverses of the eigenvalues of B_n.

For test model 1 and both methods the approximated eigenvalues accumulate in λ^* which is indicated by the continuous black line. We observe that the eigenvalues from (3.36) are all lying between the bounds for $\sigma(\mathcal{T})$ that are indicated by dashed black lines (compare Lemmas 3.2.1 and 3.2.13), while there are several eigenvalues computed from B_n that lie outside these bounds.

For test model 2 the eigenvalues obtained from (3.36) accumulate in λ_1^* and λ_2^* that are indicated by continuous lines, and they lie between the bounds for $\sigma(\mathcal{T})$ (dashed lines). However, the eigenvalues obtained from B_n don't exhibit any clear accumulation point and some of them are also lying outside the bounds for $\sigma(\mathcal{T})$. We therefore suggest to use the generalized eigenvalue problem (3.36) instead of B_n to approximate $\sigma(\mathcal{T})$. This observation for test model 2 is not very surprising since our proof of Lemma 3.3.2 relies on Assumption 3.3.1 which is not valid for test model 2.

The next step is to calculate an approximation of q (of q_1, q_2, respectively) using the approximated eigenvalues of $\mathcal{T}$. We now restrict to the eigenvalues obtained from (3.36) and compare two methods to compute the conductivity contrast. Since we clearly observed that the eigenvalues have the correct accumulation points we can compute q from

$$\lambda^* = \frac{2q}{2+q},\tag{3.37}$$

for test model 1 and q_1, q_2 from $\lambda_j^* = \frac{2q_j}{2+q_j}$ $(j = 1, 2)$ for test model 2, respectively.

For test model 1 we can also make use of the correspondence between $\sigma(\mathcal{T})$ and $\sigma(D^*)$ shown in Theorem 3.2.6. To this end we also compute the eigenvalues of a discrete version of D^* and compute q from the formula $\mu = -\frac{2+q}{2q} + \frac{1}{\lambda}$. In Section 3.3.2 we have seen that this method doesn't apply to the case of several conductivity contrasts and thus not to test model 2.

In Figure 3.5 the resulting approximations of q, q_1, q_2 are illustrated. In the plot for test model 1 we observe that the approximations using (3.37) (blue'×') and those using the correspondence between $\sigma(\mathcal{T})$ and

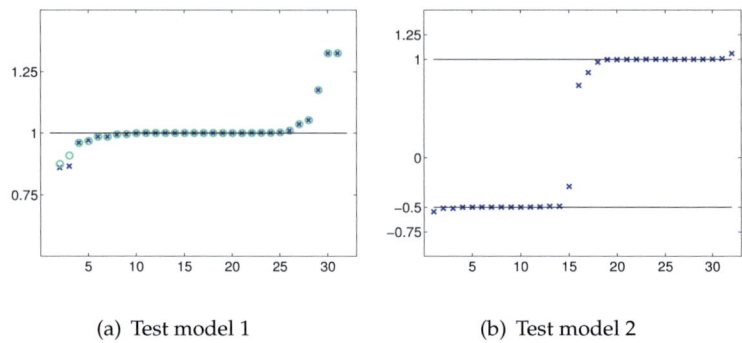

(a) Test model 1   (b) Test model 2

Figure 3.5: Approximation of q for both test models. blue '×': q (or q_1, q_2) computed from (3.37), green 'o': use of correspondence to $\sigma(D^*)$, black line: exact value of q (of q_1, q_2, respectively)

$\sigma(D^*)$ (green 'o') don't differ substantially from each other. Altogether we obtain quite accurate approximations for the exact value of q which is indicated by the continuous black line. The same holds for the second test model where the we only used the formula (3.37).

3.4.3 Inexact Data

The previous tests were conducted for unperturbed data $\Lambda_0 - \Lambda$ and under the assumption that the inclusion boundaries are known exactly. In this section we show some numerical examples for perturbed $\Lambda_0 - \Lambda$ as well as for perturbed inclusion boundaries. We start by computing the approximation of the conductivity contrasts where $\Lambda_0 - \Lambda$ is perturbed by 1% of white noise.

Figures 3.6 and 3.7 show the corresponding results for both test models. We observe that for both test models there are only few computed eigenvalues lying in the neighborhood of λ^* (of λ_1^* and λ_2^*, respectively), while the others have very large absolute values. This affects the approximations of q as well. In addition, we can only obtain an estimate of q (of q_1, q_2 respectively) from the right hand sides of the Figures 3.6, 3.7 since

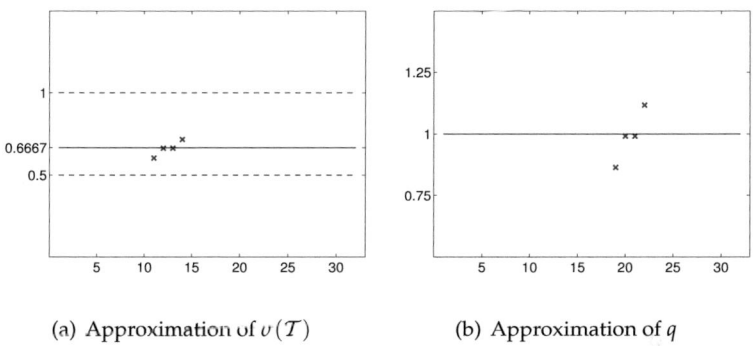

(a) Approximation of $v(\mathcal{T})$ (b) Approximation of q

Figure 3.6: Approximation of $\sigma(\mathcal{T})$ and q for test model 1 for 1% noise added to $\Lambda_0 - \Lambda$

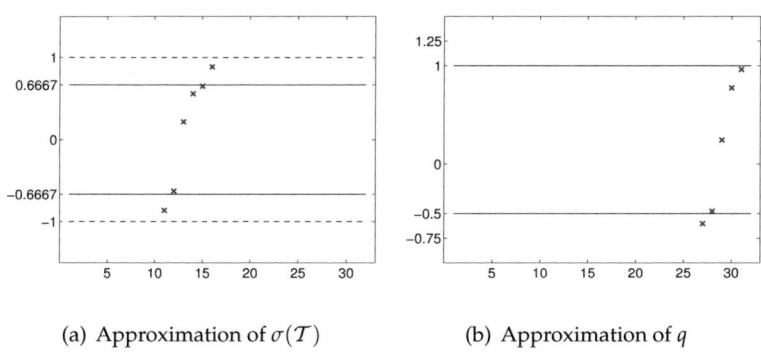

(a) Approximation of $\sigma(\mathcal{T})$ (b) Approximation of q

Figure 3.7: Approximation of $\sigma(\mathcal{T})$ and q for test model 2 for 1% noise added to $\Lambda_0 - \Lambda$

there are no distinct accumulation points as in the previous noiseless examples.

These results show that our method is quite sensitive to noise which can also be observed in the generalized eigenvalue problem (3.36): both $\Lambda_0 - \Lambda$ and $\mathcal{A}^*\mathcal{A}$ are compact operators and thus L_n and A_n are ill-conditioned. The algorithm we used to solve the generalized eigenvalue problem (3.36) performs a Cholesky factorization of A_n: $A_n = C_n C_n^\top$, where C_n is a lower triangular matrix. Now (3.36) is transformed to an ordinary eigenvalue problem by multiplication with the inverses of the ill-conditioned matrices C_n and $C_n^\top$ on both sides.

We therefore show one more example using three different methods in order to improve the numerical stability of the solution of the generalized eigenvalue problem. In the first method we we add ϵI to A_n before we perform the Cholesky factorization. The second method is inspired by Tikhonov regularization (see e.g. [50]): instead of solving (3.36) we use an eigenvalue decomposition of the operator $\left(A_n^2 + \epsilon I\right)^{-1} A_n L_n$. In the third method we conduct a spectral cut-off of A_n before solving (3.36), i.e. all eigenvalues of A_n with absolute values less than a tolerance level *tol* are set to zero. Now the matrix on the right hand side of (3.36) is singular, and the generalized eigenvalue problem is solved using the QZ algorithm (see e.g. [31]). As concrete values we used $\epsilon = 10^{-4}$ and $tol = 10^{-8}$.

The corresponding results are illustrated in Figures 3.8 and 3.9. They show that the eigenvalues obtained with $A_n + \epsilon I$ instead of A_n (red 'o') and those obtained with the spectral cut-off (pink '∇') are almost the same as the original ones (blue '$\times$'). Those obtained from the Tikhonov-like method (green '$*$') are slightly more focused around the expected accumulation points $\lambda^*, \lambda_1^*, \lambda_2^*$. However, none of the three methods yields a substantial improvement of the results for the approximated spectrum of $\mathcal{T}$ and the conductivity contrast q.

In the following test we investigate the effect of an inexactly known inclusion boundary $\partial\Omega$ on the approximations of $\sigma(\mathcal{T})$ and q. Figure 3.10 shows the exact boundary of $\partial\Omega$ (black lines) as well as the perturbed boundary (red lines).

In Figures 3.11 and 3.12 the corresponding approximations of $\sigma(\mathcal{T})$ and q are shown, where we used (3.36) without any of the three aforementioned methods to improve numerical stability. We observe that the

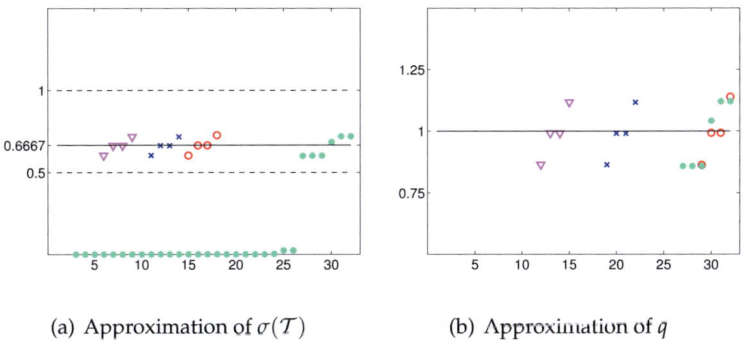

(a) Approximation of $\sigma(\mathcal{T})$ (b) Approximation of q

Figure 3.8: Approximation of $\sigma(\mathcal{T})$ and q for test model 1 for 1% noise added to L_n. blue '×': use of (3.36), red '∘': $A_n + \epsilon I$ instead of A_n in (3.36), green '∗': Tikhonov-like method, pink '∇': spectral cut-off of A_n

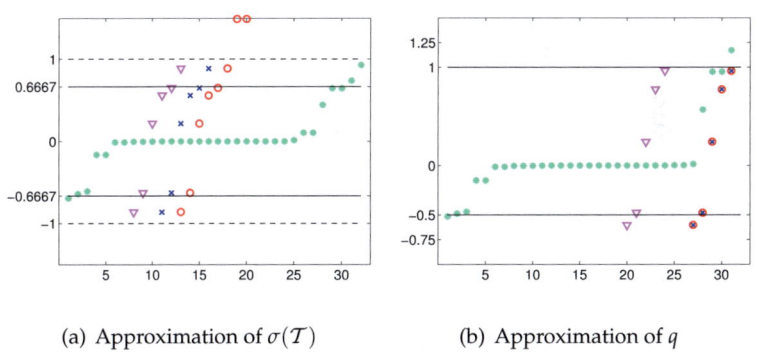

(a) Approximation of $\sigma(\mathcal{T})$ (b) Approximation of q

Figure 3.9: Approximation of $\sigma(\mathcal{T})$ and q for test model 2 for 1% noise added to L_n. blue '×': use of (3.36), red '∘': $A_n + \epsilon I$ instead of A_n in (3.36), green '∗': Tikhonov-like method, pink '∇': spectral cut-off of A_n

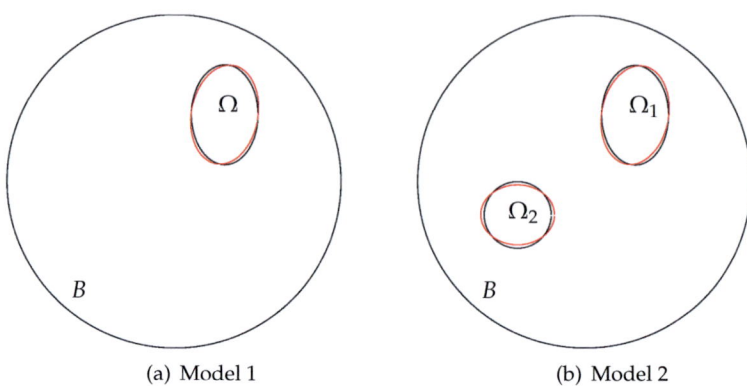

(a) Model 1   (b) Model 2

Figure 3.10: The two test models (black lines) and perturbed boundaries $\partial\Omega$ (red lines)

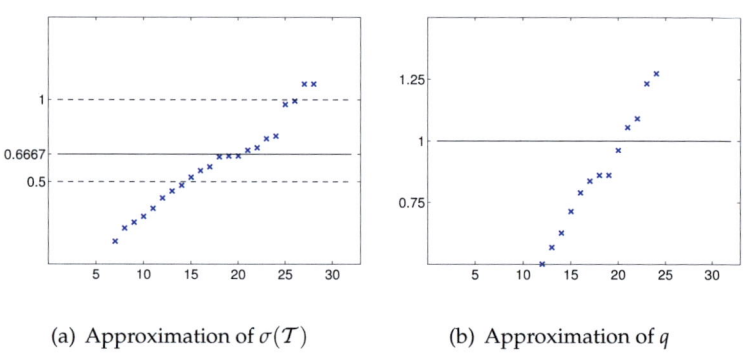

(a) Approximation of $\sigma(\mathcal{T})$ (b) Approximation of q

Figure 3.11: Approximation of $\sigma(\mathcal{T})$ and q for test model 1 and perturbed inclusion boundary $\partial\Omega$

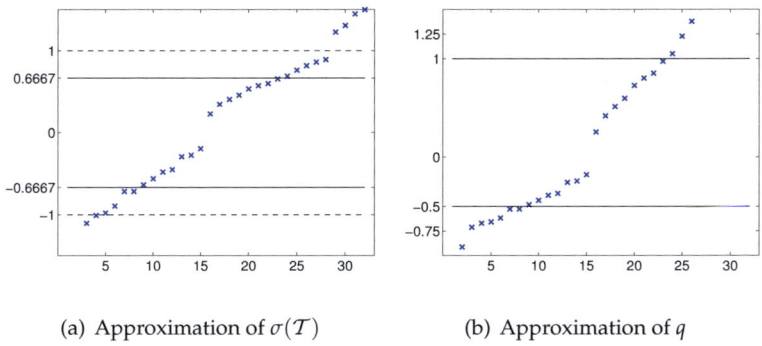

(a) Approximation of $\sigma(T)$ (b) Approximation of q

Figure 3.12: Approximation of $\sigma(T)$ and q for test model 2 and perturbed inclusion boundary $\partial\Omega$

approximated eigenvalues don't exhibit any clear accumulation points. The same holds for the approximations of q. However, for test model 1 we can read out a rough estimate of q, and for test model 2 we observe that there are two distinct values of q_1 and q_2 whose values can be read out roughly.

The above results show that our method of approximating q via the spectrum of T works quite well in the case of exact data. In the case where $\Lambda_0 - \Lambda$ is perturbed we can still obtain a rough estimate of the conductivity contrast. The same holds for the case of an imperfectly known boundary of Ω. However, in order to improve the stability of our method towards errors in the data, it would be desirable to find a reasonable regularization strategy for generalized eigenvalue problems of the type

$$Mx = \mu Nx,$$

where both M and N are ill-conditioned matrices that may be perturbed by noise as it is the case in (3.36).

3.4.4 An Alternative Approach

Since the numerical stability of our new method seems to be rather sub-optimal so far, we now present an alternative approach to determine the conductivity contrast q of a previously located inclusion Ω. This approach is based on boundary integral equation methods for a Cauchy problem.

Therefore let $f \in L^2_\diamond(\partial B)$ be an arbitrary current pattern, let $u \in H^1_\diamond(B)$ be the corresponding solution to the direct problem (1.5) and let $u_0 \in H^1_\diamond(B)$ be the solution to (3.2). Then the difference $v := u_0 - u$ solves the transmission boundary value problem (3.35).

Since we know the Neumann-to-Dirichlet operators Λ_0 and Λ, we know the corresponding boundary potential $v|_{\partial B} =: g \in L^2_\diamond(\partial B)$ for every current pattern $f \in L^2_\diamond(\partial B)$. The potential v can be represented by a single layer ansatz of the type

$$v = SL\varphi,$$

where SL denotes the single layer potential with the Neumann function as kernel as it is introduced in Section 3.2.2. In order to satisfy the Dirichlet boundary condition $v|_{\partial B} =: g$, the density $\varphi \in H_\diamond^{-\frac{1}{2}}(\partial\Omega)$ has to solve the integral equation

$$SL_{\partial\Omega \to \partial B}\, \varphi = g, \tag{3.38}$$

where the operator $SL_{\partial\Omega \to \partial B}$ is the evaluation of the single layer potential at the outer boundary ∂B. This ill-posed equation is solved using Tikhonov regularization with a fixed regularization parameter α.

After φ has been determined the conductivity contrast q can be obtained using the jump condition in the normal derivative of v at $\partial\Omega$:

$$(1+q)\, \partial_v v|_- - \partial_v v|_+ = q\, \partial_v u_0.$$

Using the jump relations for the single layer potential we deduce

$$q\left(\frac{1}{2}\varphi + D^*\varphi - \partial_v u_0\right) = -\varphi,$$

from which we obtain q e.g. by division of the maximal values of $-\varphi$ by the term in parentheses.

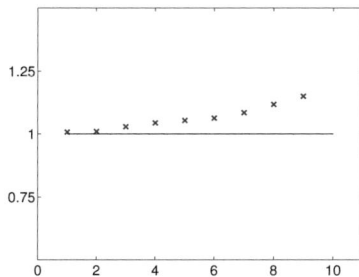

Figure 3.13: Approximated values for q for 10 different current patterns and and regularization parameter $\alpha = 10^{-6}$; exact data

We performed this procedure for test model 1 from the previous section (see Figure 3.3) for exact data $\Lambda_0 - \Lambda$ and $\partial\Omega$, for perturbed $\Lambda_0 - \Lambda$ and for an inexactly known inclusion boundary $\partial\Omega$. As a basis of current patterns we used the trigonometric basis (2.33) up to the end index $N = 5$.

The operator L_n that approximates $\Lambda_0 - \Lambda$ is computed just as in the previous section, while the integral equation (3.38) is solved on a coarser mesh.

For our first example we used exact measurement data $\Lambda_0 - \Lambda$ as well as the exact inclusion boundary $\partial\Omega$. For the regularization parameter we chose $\alpha = 10^{-6}$. Figure 3.13 shows the approximations for q, where each marker corresponds to the value obtained from one current pattern f and its corresponding image $g = (\Lambda_0 - \Lambda)f$. We observe that most of the values lie relatively close to the exact value $q = 1$, while there are some values lying further away from $q = 1$. This indicates that the approximation is strongly dependend on the specific current pattern.

In our next example the measurement data $\Lambda_0 - \Lambda$ is perturbed by 1% of white noise but we still use the exact inclusion boundary $\partial\Omega$. Figure 3.14 shows the corresponding results. We observe that we have to use the larger regularization parameter $\alpha = 10^{-5}$ and that the approximated values for q differ slightly more from $q = 1$ than for the case of exact

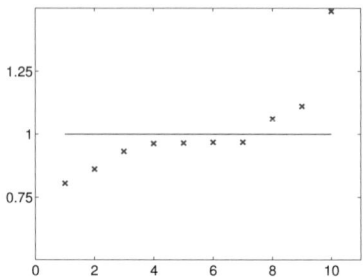

Figure 3.14: Approximated values for q for 10 different current patterns and and regularization parameter $\alpha = 10^{-5}$; 1% white noise added to $\Lambda_0 - \Lambda$

measurement data. However, the approximations are noticeably better than those in Figure 3.6 where we there were only few values lying in the neighborhood of q.

Or last example deals with an imperfectly known inclusion boundary $\partial\Omega$, and we used the example with rotated boundary as it is illustrated in Figure 3.10. The results are shown in Figure 3.15. Again, they are noticeably worse than for the exact data case but still better than the results from Figure 3.11.

These observations indicate that the method to obtain the value of q that is presented in this section might be an alternative to our new method from the previous sections. Our numerical examples only serve as a proof of concept, and one has to conduct a much wider range of tests in order to evaluate the potential of both methods.

However, since there is a variety of regularization techniques for the solution of an integral equation of the first kind as in (3.38), numerical stability towards perturbations in the data can be established for the present method. In contrast, our new method from the previous sections is based on the solution of the generalized eigenvalue problem (3.36) with ill-conditioned operators on both sides for which we have no regularization strategies so far.

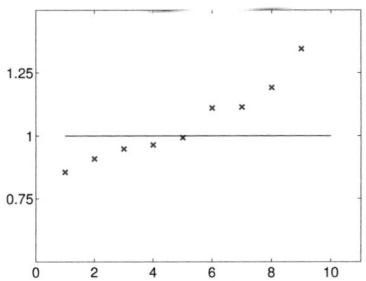

Figure 3.15: Approximated values for q for 10 different current patterns and and regularization parameter $\alpha = 10^{-6}$; perturbed inclusion boundary $\partial\Omega$

4 Conclusions

The considerations in Chapter 2 show that our modified version of the Factorization method for EIT is capable of detecting inclusions in the mixed case, i.e. the case in which there are inclusions with a lower absolute conductivity than the background as well as inclusions with a higher absolute conductivity than the background. Furthermore, we showed that the problem setting can even be extended to insulating and perfectly conducting inclusions.

In our numerical tests we showed that using this new method we can detect both inclusion types using a two-step algorithm. However, we observed that the reconstruction quality is dependent on the choice of the parameters ρ_1, ρ_2 as well as the size of the covering domains $\tilde{\Omega}_1, \tilde{\Omega}_2$. We therefore suggest to choose the absolute values of ρ_1, ρ_2 very small compared to the largest eigenvalues of the operator $|\mathrm{Re}\,\Lambda - \Lambda_0| + \mathrm{Im}\,\Lambda$. In addition, we suggest to choose the covering domains $\tilde{\Omega}_1, \tilde{\Omega}_2$ as small as it is admitted by the available a priori information about the inclusions' locations.

The comparisons to the original Factorization method for EIT showed that the reconstructions obtained with the original Factorization method are noticeably better than those obtained with our new method. However, if the parameters and the covering domains in our new method are chosen reasonably as described above, then it is also capable of reconstructing both types of inclusions for exact data as well for noisy data. Nevertheless, a proof that the original Factorization method works even in the mixed case would be the optimal solution to this problem. However, to our knowledge, there is no such proof so far.

In future work one could also introduce a third type of inclusions which involves conductivities with strictly negative imaginary parts and extend the covering method to these.

Moreover, in Chapter 3 we presented a new version of the Factorization method for EIT that is based on the factorization of the operator $\Lambda_0 - \Lambda$ in three operators that are different from those used in e.g. Chapter 2. The main advantage of this new factorization is that the middle operator is a bounded linear operator from the space $L^2(\Omega, \mathbb{R}^d)$ to itself which implies that its spectrum is a compact set in $\mathbb{R}$ (or in $\mathbb{C}$, respectively).

We showed that this spectrum is closely related to the conductivity inside the inclusions and presented a method to compute this spectrum and the conductivity contrast numerically. The numerical experiments with this method show that it works quite well for exact data but lacks numerical stability for perturbed data. However, this observation is not surprising since the determination of the conductivity contrast from the knowledge of the Neumann-to-Dirichlet map is still an ill-posed problem.

The main reason for this instability lies in the generalized eigenvalue problem (3.29) that has compact operators on both sides. We tried to apply some regularization strategies to the operator on the right hand side of this generalized eigenvalue problem which didn't improve the results substantially. It would be therefore desirable to have regularization strategies for such generalized eigenvalues.

In Section 3.4.4 we presented an alternative method to determine the conductivity contrast that is based on a boundary integral equation of the first kind. To this integral equation we can apply well-known regularization strategies to handle noisy data.

Our considerations in Chapter 3 are based on the assumption of piecewise constant conductivities. In future work one could try to extend this assumption to e.g. piecewise continuous conductivities and investigate the structure of the spectrum of the middle operator in this case. In addition, another obvious extension of this work is to apply the covering method from Chapter 2 to the new factorization from Chapter 3 since for the new factorization the middle operator also fails to be coercive in the case of mixed inclusions.

List of Figures

List of Symbols

$H_\diamond^{\frac{1}{2}}(\partial B)$ .... space of mean-free functions in $H^{\frac{1}{2}}(\partial B)$, page 4

$\tilde{\Lambda}_j$ .... modification of $\Lambda - \Lambda_0$, page 38

$\tilde{\Omega}_j$ .... covering domain, page 30

$\tilde{G}^{(j)}$ .... auxiliary operator, page 34

$\tilde{G}_j$ .... auxiliary operators, page 34

$\tilde{Q}_j$ .... auxiliary operators, page 31

φ_y .... dipole potential, page 23

A_n .... discrete version of $\mathcal{A}^*\mathcal{A}$, page 105

B .... investigated subject, page 5

B_n .... discrete version of $\mathcal{B}\mathcal{B}$, page 105

$C_\diamond^2(\partial B)$ .... space of mean-free C^2-functions, page 4

D, D^* .... boundary integral operators, page 74

D_j^* .... integral operator D^* on $\partial\Omega_j$, page 85

$D_{j\to k}^*$ .... integral op. D^* on $\partial\Omega_j$ evaluated on $\partial\Omega_k$, page 85

DL .... double layer potential, page 72

f .... current pattern, page 5

G .... operator in factorization, page 15

$H^1(B)$ .... Sobolev space of order 1, page 5

$H^{-\frac{1}{2}}(\partial B)$ .... Sobolev space of order $-\frac{1}{2}$, page 4

$H^{\frac{1}{2}}(\partial B)$ .... Sobolev space of order $\frac{1}{2}$, page 4

$L^2(B, \mathbb{R}^d)$ .... space of L^2-vector fields in B, page 4

$L^\infty(B)$ .... space of essentially bounded functions, page 5

$L_\diamond^2(\partial B)$ .... space of mean-free L^2-functions, page 4

L_n .... discrete version of $\Lambda_0 - \Lambda$, page 105

N .... Neumann function, page 23

P_n, Q_n .... orthogonal projections, page 98

q .... conductivity contrast, page 62

$Q^{(j)}$ .... auxiliary operator, page 31

q_j .... conductivity contrast, page 62

S .... boundary integral operators, page 74

S_j .... middle operator in modified factorization, page 38

SL .... single layer potential, page 72

SL_j .... single layer potential on $\partial\Omega_j$, page 85

T .... operator in factorization, page 16

$T^{(ij)}$ .... partial operator of T, page 24

Index

Bibliography

[1] G. ALESSANDRINI, *Stable determination of conductivity by boundary measurements*, Appl. Anal., 27 (1988), pp. 153–172.

[2] ——, *Unique determination of multiple cracks by two measurements*, SIAM J. Control Optim., 34 (1996), pp. 913–921.

[3] T. ARENS AND N. I. GRINBERG, *A complete factorization method for scattering by periodic structures*, Computing, 75 (2005), pp. 111–132.

[4] T. ARENS AND A. KIRSCH, *The factorization method in inverse scattering from periodic structures*, Inverse Problems, 19 (2003), pp. 1195–1211.

[5] K. ASTALA AND L. PÄIVÄRINTA, *Calderón's inverse conductivity problem in the plane*, Annals of Mathematics, 163 (2006), pp. 265–299.

[6] L. BORCEA, *Electrical impedance tomography*, Inverse Problems, 18 (2002), pp. R99–R136.

[7] R. M. BROWN AND R. H. TORRES, *Uniqueness in the inverse conductivity problem for conductivities with 3/2 derivatives in L^p, $p > 2n$*, J. Fourier. Anal. Appl., 9 (2003), pp. 1049–1056.

[8] M. BRÜHL, *Gebietserkennung in der elektrischen Impedanztomographie*, PhD thesis, Universität Karlsruhe, Karlsruhe, Germany, 1999.

[9] ——, *Explicit characterization of inclusions in electrical impedance tomography*, SIAM J. Math. Anal., 32 (2001), pp. 1327–1341.

[10] M. BRÜHL AND M. HANKE, *Numerical implementation of two noniterative methods for locating inclusions by impedance tomography*, Inverse Problems, 16 (2000), pp. 1029–1042.

[11] M. BRÜHL, M. HANKE, AND M. PIDCOCK, *Crack detection using electrostatic measurements*, Math. Model. Numer. Anal., 35 (2001), pp. 595–605.

[12] F. CAKONI AND D. COLTON, *The determination of the surface impedance of a partially coated obstacle from far field data*, SIAM J. Appl. Math., 64 (2004), pp. 709 – 723.

[13] F. CAKONI, D. COLTON, AND H. HADDAR, *The computation of lower bounds for the norm of the index of refraction in an anisotrpic media from far field data*, Journal of Integral Equations and Applications, 21 (2009), pp. 203 – 227.

[14] F. CAKONI, D. COLTON, AND P. MONK, *On the use of transmission eigenvalues to estimate the index of refraction from far field data*, Inverse Problems, 23 (2007), pp. 507–522.

[15] A. CALDERÓN, *On an inverse boundary value problem*, in Seminar on Numerical Analysis and its Applications to Continuum Mechanics, Rio de Janerio, 1980, Soc. Brasileira de Matemática, pp. 65–73.

[16] M. CHENEY, D. ISAACSON, AND J. NEWELL, *Electrical impedance tomography*, SIAM Review, 41 (1999), pp. 85–101.

[17] M. CHENEY, D. ISAACSON, J. NEWELL, S. SIMSKE, AND J. GOBLE, *Noser: an algorithm for solving the inverse conductivity problem*, Int. J. Imag. Syst. Technol., 2 (1990), pp. 66–75.

[18] K.-S. CHENG, D. ISAACSON, J. NEWELL, AND D. GISSER, *Electrode models for electric current computed tomography*, IEEE Trans. Biomed. Engrg., 36 (1989), pp. 918–924.

[19] D. COLTON AND R. KRESS, *Integral Equation Methods in Scattering Theory*, Wiley-Interscience, 1983.

[20] ——, *Inverse acoustic and electromagnetic scattering theory*, Springer, 2nd ed., 1998.

[21] R. DAUTRAY AND J. L. LIONS, *Functional and variational methods*, vol. 2 of Mathematical analysis and numerical methods for science and technology, Springer, 1988.

[22] D. C. DOBSON, *Convergence of a reconstruction method*, SIAM J. Appl. Math., 52 (1992), pp. 442–458.

[23] H. ECKEL AND R. KRESS, *Nonlinear integral equations for the inverse electrical impedance problem*, Inverse Problems, 23 (2007), pp. 475–491.

[24] E. FRANCINI, *Recovering a complex coefficient in a planar domain from the Dirichlet-to-Neumann map*, Inverse Problems, 16 (2000), pp. 107–119.

[25] B. GEBAUER, *The factorization method for real elliptic problems*, Z. Anal. Anwend, 25 (2006), pp. 81–102.

[26] ——, *Localized potentials in electrical impedance tomography*, Inverse Problems and Imaging, 2 (2008), pp. 251–269.

[27] B. GEBAUER AND N. HYVÖNEN, *Factorization method and irregular inclusions in electrical impedance tomography*, Inverse Problems, 23 (2007), pp. 2159–2170.

[28] ——, *Factorization method and inclusions of mixed type in an inverse elliptic boundary value problem*, Inverse Problems and Imaging, 2 (2008), pp. 355–372.

[29] D. GILBARG AND N. TRUDINGER, *Elliptic Partial Differential Equations of Second Order*, Springer, New York etc., 1983.

[30] D. GISSER, D. ISAACSON, AND J. NEWELL, *Electric current computed tomography and eigenvalues*, SIAM J. Appl. Math., 50 (1990), pp. 1623–1634.

[31] C. F. GOLUB, GENE H. ; VAN LOAN, *Matrix computations*, Johns Hopkins series in the mathematical sciences, Johns Hopkins Univ. Pr., 1996.

[32] R. GRIESMAIER, *Reconstruction of thin tubular inclusions in three-dimensional domains using electrical impedance tomography*, SIAM J. Imaging Sci., 3 (2010), pp. 340–362.

[33] N. I. GRINBERG, *Obstacle visualization via the factorization method for the mixed boundary value problem*, Inverse Problems, 18 (2002), pp. 1687–1704.

[34] ——, *On the inverse obstacle scattering problem with robin or mixed boundary condition: application of the modified Kirsch factorization method*, tech. rep., 2002.

[35] ——, *The factorization method in inverse obstacle scattering*, habilitation thesis, University of Karlsruhe, Karlsruhe, 2004.

[36] ——, *The operator factorization method in inverse obstacle scattering*, Integr. equ. oper. theory, 54 (2006), pp. 333–348.

[37] N. I. GRINBERG AND A. KIRSCH, *The factorization method for obstacles with a-priori separated sound-soft and sound-hard parts*, Mathematics and Computers in Simulation, 66 (2004), pp. 267–279.

[38] H. HADDAR, S. KUSIAK, AND J. SYLVESTER, *The convex backscattering support*, SIAM J. Appl. Math., 66 (2006), pp. 591–615.

[39] M. HANKE AND M. BRÜHL, *Recent progress in electrical impedance tomography*, Inverse Problems, 19 (2003), pp. S65–S90.

[40] M. HANKE, N. HYVÖNEN, AND S. REUSSWIG, *Convex source support and its application to electric impedance tomography*, SIAM Journal on Imaging Sciences, 1 (2008), pp. 364–378.

[41] ——, *An inverse backscatter problem for electric impedance tomography*, SIAM Journal on Mathematical Analysis, 41 (2009), pp. 1948–1966.

[42] L. HARHANEN AND N. HYVÖNEN, *Convex source support in half-plane*, Inverse Problems and Imaging, 4 (2010), pp. 429–448.

[43] F. HETTLICH AND W. RUNDELL, *The determination of a discontinuity in a conductivity from a single boundary value measurement*, Inverse Problems, 14 (1998), pp. 931–947.

[44] N. HYVÖNEN, *Complete electrode model of electrical impedance tomography: approximation properties and characterization of inclusions*, SIAM J. Appl. Math., 64 (2004), pp. 902–931.

[45] ——, *Application of the factorization method to the characterization of weak inclusions in electrical impedance tomography*, Adv. in Appl. Math., 39 (2007), pp. 197–221.

[46] N. HYVÖNEN, H. HAKULA, AND S. PURSIAINEN, *Numerical implementation of the factorization method within the complete electrode model of impedance tomography*, Inverse Problems and Imaging, 1 (2007), pp. 299–317.

[47] V. ISAKOV, *Inverse Problems for Partial Differential Equations*, Springer, 1998.

[48] T. KATO, *Perturbation theory for linear operators*, Springer, repr. of the 1980 ed., 1995.

[49] A. KIRSCH, *Generalized boundary value- and control problems for the Helmholtz equation*, 1984.

[50] ——, *An introduction to the mathematical theory of inverse problems*, Springer, 1996.

[51] ——, *Characterization of the shape of a scattering obstacle using the spectral data of the far field operator*, Inverse Problems, 14 (1998), pp. 1489–1512.

[52] ——, *Factorization of the far field operator for the inhomogeneous medium case and an application in inverse scattering theory*, Inverse Problems, 15 (1999), pp. 413–429.

[53] ——, *The factorization method for a class of inverse elliptic problems*, Math. Nachr., 278 (2004), pp. 258–277.

[54] ——, *The factorization method for Maxwell's equations*, Inverse Problems, 20 (2004), pp. S117–S134.

[55] ——, *Inverse Probleme WS08/09.* lecture notes, 2009.

[56] A. KIRSCH AND N. I. GRINBERG, *The Factorization Method for Inverse Problems*, Oxford Lecture Series in Mathematics and its Applications 36, Oxford University Press, 2008.

[57] K. KNUDSEN, M. LASSAS, J. L. MUELLER, AND S. SILTANEN, *Regularized d-bar method for the inverse conductivity problem*, Inverse Problems and Imaging, 3 (2009), pp. 599–624.

[58] R. KOHN AND M. VOGELIUS, *Determining conductivity by boundary measurements*, Comm. Pure Appl. Math., 37 (1984), pp. 113–123.

[59] R. KRESS AND G. ROACH, *On the convergence of successive approximations for an integral equation in a green's function approach to the dirichlet problem*, Journal of mathematical analysis and applications, 55 (1976), pp. 102–111.

[60] S. KUSIAK AND J. SYLVESTER, *The scattering support*, Commun. Pure Appl. Math., 56 (2003), pp. 1525–1548.

[61] A. LECHLEITER, *A regularization technique for the factorization method*, Inverse Problems, 22 (2006), pp. 1605–1625.

[62] ———, *Factorization Methods for Photonics and Rough Surface Scattering*, PhD thesis, Universität Karlsruhe, Karlsruhe, Germany, 2008.

[63] ———, *The Factorization method is independent of transmission eigenvalues*, Inverse Problems and Imaging, 3 (2009), pp. 123–138.

[64] A. LECHLEITER, N. HYVÖNEN, AND H. HAKULA, *The factorization method applied to the complete electrode model of impedance tomography*, SIAM J. Appl. Math., 68 (2008), pp. 1097–1121.

[65] A. LECHLEITER AND A. RIEDER, *Newton regularizations for impedance tomography: convergence by local injectivity*, Inverse Problems, 24 (2008), p. 065009 (18pp).

[66] W. MCLEAN, *Strongly Elliptic Systems and Boundary Integral Operators*, Cambridge University Press, Cambridge, UK, 2000.

[67] C. MIRANDA, *Partial differential equations of elliptic type*, Springer, 2., rev. ed., 1970.

[68] A. NACHMAN, *Global uniqueness for a two-dimensional inverse boundary value problem*, Ann. of Math., 142 (1995), pp. 71–96.

[69] M. REED AND B. SIMON, *Methods of modern mathematical physics*, vol. 1: Functional analysis, Acad. Pr., New York, 1972.

[70] A. RIEDER, *Keine Probleme mit Inversen Problemen*, Vieweg, 1. ed., 2003.

[71] S. SCHMITT, *The factorization method for EIT in the case of mixed inclusions*, Inverse Problems, 25 (2009), p. 065012 (20pp).

[72] E. SOMERSALO, M. CHENEY, AND D. ISAACSON, *Existence and uniqueness for electrode models for electric current computed tomography*, SIAM J. Appl. Math., 52 (1992), pp. 1023–1040.

[73] J. SYLVESTER AND G. UHLMANN, *A global uniqueness theorem for an inverse boundary value problem*, Ann. Math., 125 (1987), pp. 69–153.

[74] E. SÜLI AND D. F. MAYERS, *An introduction to numerical analysis*, Cambridge University Press, Cambridge, 1. publ. ed., 2003.

[75] G. UHLMANN, *Electrical impedance tomography and Calderon's problem*, Inverse Problems, 25 (2009), p. 1230011 (39pp).